Why is there a frog in my throat?

A Guide to Hoarseness

153/500
9 Aug 2013
Your voice!

JAMES P. THOMAS, MD
Portland, Oregon, USA

First Edition, December 2012

edition 1.03 November 2012
ISBN: 978-0-9855701-0-1
Library of Congress Control Number: 2012911391

Book written and designed by James P. Thomas, MD
Printed in the United States of America

Thanks to Jody Coale, RN for ongoing feedback of the manuscript and to my
patients and my fellows who continue to teach me about the voice.
                    - James P. Thomas, MD

Dedicated to Michael Teixido, MD & Robert Bastian, MD

Michael
who serendipitously alters my life from time to time;

& Robert
who taught me how to hear and see the unnoticed

Contents

Why is there a frog in my throat?

Introduction

What to expect from this book

When something goes wrong with your voice, you might simply say, "I am hoarse." Your funny best friend might ask with a smile, "Got a frog in your throat?" Then what? Do you rest your voice? Do you gargle? Do you eat chicken soup? Do you see a doctor? Do you pick up a pill at the pharmacy? Which doctor do you see?

What if you are a doctor? Is it simple or difficult to discover the cause of a hoarse voice? What do you need to do to make a diagnosis? Do you look in the mouth? Do you guess? Do you write a prescription for a pill?

Unfortunately, after saying, "I am hoarse," getting a diagnosis can be a long, tough process. You may see your family doctor or even several specialists. You may receive some type of treatment. You may be told that it will get better with time. In any case, you may wonder about the explanations you receive.

If you are saying, "I am hoarse," you have every right to ask your doctor not only what you should do, but, "Why am I hoarse?" Understanding why you are hoarse will more likely put you on, and keep you on, an optimal treatment path.

As a doctor, if you have been guessing, or just passively doing what you were taught, what can you do to make your guess more educated? Are your patients coming back, but not really getting better as often as you would like? Can you really see the cause of hoarseness? Have you ever wondered if there really is a purple pill that can cure hoarseness?

I purchased an endoscope and a stroboscope and began filming vocal cords, trying to find something wrong with hoarse voices, finding a visible problem at times, but not consistently. Every month or two I took a week off from my otolaryngology practice to study with various laryngologists, seeing what I could learn and improve my skills. I visited one laryngologist in Chicago, Robert Bastian, and discovered a physician who had taught himself how to listen to a patient's history and her voice, then describe what we were likely to find before we looked in the throat.

I returned the following year to study with Robert for another six months. After that study, I limited my practice to only laryngeal disorders and soon became aware that I was able to hear voice problems before I made a video recording of the vocal cords with an endoscope. That was the beginning of when I felt like I was becoming a voicedoctor, when I could both hear and see voice disorders.

On my website www.voicedoctor.net, I present some of my ideas about voice disorders that with regularity trigger e-mail from people with a voice problem. My patients tell me strikingly similar stories in the office. While each person's story is their own, frequent themes are:

- The explanation for my hoarseness doesn't make sense.

- I have tried so many treatments without success.

When I put a tiny camera into the throat and show a patient her own vocal cords enlarged on a video screen, I have the sense that the views of the moving vocal cords, coupled with an understandable explanation, often meet the needs of the hoarse patient. Frequently enough, the explanation leads further, to a remedy or a resolution, though at a minimum, it gives at least an understanding by the patient of what her voice problem is. The explanation for the hoarseness makes sense. This understanding seems to be more than she has received before and consequently, I write this book so that more people might come to understand hoarseness.

Perhaps you already have some idea of what hoarseness is, but can't further define it. Perhaps you can hear a hoarseness, but can't quite pin down the cause of it. You, doctor or patient, may find something of value in this book.

In Part I, voice is defined and I sort out what we really mean when we say, "My voice is hoarse." In effect, I define the difference between normal voice and abnormal voice. This will lay a straightforward foundation for how to think about hoarseness.

Part II categorizes and maps out specific types of vocal problems. This approach makes problems with the voice easier to identify. A correctly identified voice problem is really the only reasonable way to start an effective treatment.

I receive many, many emails that begin with, "I have been diagnosed with _____ as a cause of my hoarseness ..." and they conclude with, "but I am not getting any better. Can you suggest a stronger pill to treat _____?"

Imagine, before you ask for a stronger pill, would you prefer to have an accurate diagnosis of an infection that is then treated with an *average antibiotic* or would you rather have the *most powerful antibiotic* in the world, for a problem which turns out not to be an infection? The accuracy of the diagnosis matters more than power of the treatment. So, a hoarse person with a voice that is not improving is far more likely to be suffering from an inaccurate diagnosis, than from inadequate strength of medication. Therefore I cannot suggest a stronger pill when the assistance I can most likely offer, and really must offer, is a more accurate diagnosis.

The categorization of thinking described in this book has been helpful to me for years in my laryngeal medical practice. Even as I intermittently encounter vocal problems I have never seen before, these new problems still seem to fit into one of these categories. I continue to gain confidence in this categorization.

Part III is a bit more technical and highlights various ways of examining the voice. Perhaps you are a student or a new physician. Perhaps you don't have much equipment or have never been taught how to use endoscopy equipment to its full potential for examining the voice. Quite possibly you are a patient or even a primary care physician and don't have any equipment at all. I highlight how you might determine if you are getting a good examination from an oto-laryngologist (also known as an ENT in the US or an ORL in other parts of the world), or even a subspecialist known as a <u>laryngologist</u>[1].

Part IV discusses thinking errors (e.g. if everyone believes in a pill, it must work). I also discuss how to maximize the value of an examination. If you are hoarse, almost certainly you will benefit from a video recording of your vocal cords. Most vocal events happen too fast for you to perceive, without some capability to slow and review the images of a video recording. Even video needs some help from a stroboscope or high speed recording to catch events well. If you are able to view a video of your vocal cords moving, along with the framework described in this book for understanding normal and abnormal voice, you might be able to identify the probable location of your hoarseness. At a minimum, you should be able to estimate whether or not your physician was directing you in an appropriate direction.

Hoarseness can be visualized and with a few guidelines can be understood. This book will provide a framework for that understanding.

All of the stories in this book are true cases. For privacy, all of the patient names are fictional. While the gender of the patient is unchanged in each case described, I try to randomly use male and female pronouns when referring to generic people or physicians.

---

1 Physicians who specialize in voice are known around the world by various names: laryngologist, phonosurgeon, voice doctor, phoniatrist and there are probably others. Essentially these words mean that the person has a special interest in the voice beyond what a general doctor has; beyond even what a typical ear, nose and throat physician has. They are focused to some degree on the voice. There is not a formal degree for a laryngologist and anyone may use the name. Quite likely, over time the training for this subspecialty will become more formalized.

Precision

Diagnosis is the foundation that treatment is built on.

There are a great many skilled physicians and there are a great many wonderful drugs. Surgeries are becoming more and more delicate. Physician skills and pharmaceuticals have improved tremendously during my short career as a physician. Yet a precise and accurate diagnosis can still be difficult to come by. For one thing, a precise diagnosis takes time. A second issue is that a precise diagnosis takes as much skill, albeit a different skill, as surgery. Otolaryngology, the speciality most closely associated with the voice, is traditionally a surgical subspecialty. In surgical training programs, surgery skills are typically emphasized over diagnostic skills.

However, without the foundation of a precise diagnosis, the best surgeon in the world gets random results. Without a precise and accurate diagnosis, the strongest drug in the world ends up treating the wrong condition. So wonderful drugs and wonderful surgeons require a diagnostician who is both precise and accurate.

In *The Black Swan*, Nassim Nicholas Taleb, a persuasive author of numbers, probability and specifically the difficulty of understanding what you have never seen, applies his expertise on probability to his own personal experience with medicine[2]. An expert will almost always know more than you in the area of their expertise. He advises, "No matter what anyone tells you, it is a good idea to question *the error rate* of an expert's procedure. Do not question his procedure, only his confidence."

Most surgeons are reasonably technically adept because they were selected for their manual skills. The training programs are well designed to select for excellent surgical skills (though there is still variation). Surgeons tend to enjoy performing surgery and typically

---

2   Nassim Nicholas Taleb, *The Black Swan: The Impact of the Highly Improbable*, (New York: Random House, 2007), 145.

perform surgery frequently after their training, gaining a great deal of confidence. Still, even surgeons seem to have a difficult time assessing their own error rate, especially their rate of missed diagnosis.

So even if you are willing to faithfully believe in your surgeon's technical skills (based on his description to you of his own technical prowess), how can you assess his diagnostic skills and his error rate, when even he cannot? When I am a patient, rather than ask a physician about his error rate, about the closest I can come to assessing a physician's diagnostic skill prior to an intervention is to ask myself several less direct questions:

- Do I understand the physician's explanation of my problem?

- Do I understand why the intervention should correct the problem?

- Do I understand what my physician is uncertain about?

For example, I can ask my physician, what will taking this pill specifically accomplish? What will abstaining from caffeine specifically change about my symptoms? What are the trade-offs being taken in a decision to have surgery? (There are always trade-offs!)

If you understand his explanation, it is more likely to be a correct explanation. If he makes predictions about your condition that you can test over time, then you can assess the accuracy of his confidence. In laryngology you should be able to look at the video recording of your vocal cords and see if what your doctor says is making sense. The function of the larynx is simple in that it functions very much like a mechanical machine. It adheres to the laws of physics. It is a valve that opens and closes. It is a pair of strings that vibrate. Consequently, most people can understand the mechanics of the vocal cords. An accurate and precise diagnosis can be translated into easy to understand, non-medical terms and is the key to treatment.

Another caveat, if you feel belittled by your doctor for questioning his judgment, beware. If you have to take the explanation of your physician on the basis of faith, telling yourself that you are not as smart as your physician, consider the possibility that the wool is being pulled over your eyes. Curiously, in some instances a doctor bluffs, hiding behind the facade of expertise.

Admittedly it is very difficult for anyone to know what he doesn't know, even for the very bright people who are often selected for medical school. When a physician acknowledges what he doesn't know, he tells the patient about this gap in knowledge or he refers the patient to someone else.

So, let's say you have a hoarse voice and want to know more. Read on.

Part I – Laryngology definitions

Voice

What is voice?

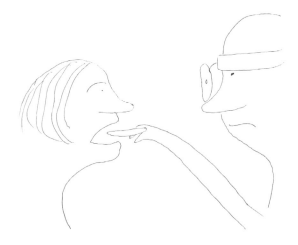

Faith N. Metsan felt something begin to change in her throat one autumn afternoon. Over the next few hours, her voice just seemed to disappear. She didn't have any other symptoms. By evening, when she tried to speak, her voice was nothing more than a whisper. Thinking this was just the beginning of laryngitis, and because work was busy, she waited for a week. When she didn't seem to be getting better she called her primary care physician, Dr. Marcus Goodew, and made an appointment.

At Dr. Goodew's office, his nurse recorded Faith's blood pressure and vital signs. Then Dr. Goodew listened to her heart and lungs.

He palpated her neck for swollen glands, looked into her ears, and of course, into her mouth. He asked her to open her mouth, say "ah" and lastly peered in with a small light.

He did not look at her vocal cords, either resting or vibrating. How can we infer that? The vocal cords are located around a bend in the throat. At a minimum, it takes an angled mirror, placed in the back of the mouth to see the vocal cords. Even then, the view can be quite fleeting. New technologies such as endoscopes that go through the mouth or the nose allow a more prolonged and direct view. Faith doesn't recall anything being put in her nose or her mouth.

If Faith's problem is with her voice, we should reasonably ask, "What is voice?" I have heard answers such as "voice is how we communicate" or "sounds that come out of our throat." Some people have said, "the sounds our vocal cords make." Let's digress from her story to think about what the word voice really means.

Consider that a violin has a "voice." A bow is pulled across the strings, they vibrate and the air inside the wood of the violin resonates, amplifying the vibrations. We hear and recognize the sound of the violin. We can say that we recognize the voice of the violin.

Effectively, any thing that vibrates in the audible range of humans has a voice, whether the horn in a car or lips placed against a trumpet's mouthpiece. Voice is a sound produced by vibration and amplified by resonance; no vocal cords are required.

We can tune a wire to vibrate 150 times per second and we will hear a tone. On a piano, the white key at D3 vibrates at almost 150 times a second, creating an audible vibration. The sound board of a particular piano resonates and amplifies the pitch, bringing into play various overtones. The surrounding strings are induced to vibrate subtly and add additional harmonics. For those of us not used to listening closely, we may not initially be able to describe in words the voice of this particular piano. Yet, play the same note on a guitar string and we hear the same tone, but a different voice. And most of us could separate the voice of a guitar from a piano.

A piano enthusiast who has spent thousands of hours listening to various pianos may be able to detect the different voice belonging to a Bosendorfer piano as compared to a Mason and Hamlin. While many of us could not distinguish the voice difference between pianos, because we have spent thousands of hours listening to other humans and to ourselves, when there is a slight change in the quality of our voice or a friend's voice, we can usually detect it relatively quickly.

So voice consists of a sound source, that is, something vibrating (usually vocal cords in a human) and an amplifier, a resonating cavity. Faith's complaint is that she cannot produce much of a sound, so something is wrong with the source of her vibrations. Her vocal cords presumably did not evaporate on that autumn afternoon, so her problem is with an altered or diminished vibration of her vocal cords.

Later we will learn how Dr. Goodew could see and identify diminished or altered vibrations (for technical details visit "Seeing the vocal cords" on page 261). For now, let's consider what is not wrong and where we should not be looking when we examine Faith by considering another story.

Speech

How is speech different from voice?

Mr. Heim Stillear, at 70 years old, seems older and less intelligent to his family, since he was hospitalized for a stroke about two months ago. They come to the appointment with him and help him into the exam chair. When I ask Heim what is going on with him, the long pause before his strained answer begins makes his family uncomfortable. His daughter interjects, "We are having a very difficult time understanding him." I learn that the right side of his body is still not completely under his control. He is on blood thinners to prevent another blood clot from floating though his system, plugging up another part of his brain. In rehabilitation, physical therapists teach him to walk and speech therapists coach him to improve his swallowing and his speaking.

When I address him again, and he responds, his voice is loud enough, but his words are slurred and unclear. I notice that his lips at the side of his mouth droop. His smile is out of kilter. I look further: one side of his tongue has little writhing movements (fasciculations) and when he sticks his tongue out, it always moves off toward one side.

Some people ask, aren't voice and speech the same thing? While they are related, they are not the same. However, in practice, speech problems and voice problems can be easily confused. Let's distinguish them.

An easy mental image we can use to separate speech from voice is to draw a line, which we will call the "Speech Line," across the neck above the Adam's apple. This Speech Line roughly separates speech production above from voice production below. The two systems interact and to some degree overlap. Yet in general, below the Speech Line, sound is created, above the Speech Line, sound is modified into language.

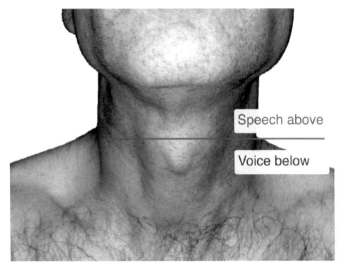

Speech above

Voice below

Below the Speech Line, the larynx produces audible vibrations – sound – in the normal human system. Above the Speech Line sound is modified. More specifically, vowels are the modification of the resonance cavities (mostly the pharynx and tongue modify the shape of the resonating cavities) and consonants are the interruptions or restrictions of the airflow. The interrupters include the palate, the tongue and the lips. These alterations in the airflow, when combined, coalesce to form words, then phrases, then sentences and we begin to communicate. We achieve speech.

Since Mr. Stillear's problem is with difficulty producing *words* clearly, the issue is most likely in the upper half of this system. A physician would call this dysarthria. His problem is not with sound production because his volume is good, and any single sound is quite clear on its own. When I look at his vocal cords, they are indeed vibrating well.

However, I focus my examination above the Speech Line: on his face he has difficulty moving his lips; in his mouth he has difficulty controlling the movement of his tongue. These motion impairments garble his speech. In adults, most new problems with speech are neurologic problems. His stroke has impaired the neural input to his tongue and his lips so his speech lacks clarity. It is even possible that his intelligence is entirely unaffected by his stroke. The slurring merely gives his family the perception of diminished intelligence.

For the most part, I am not going to cover speech disorders in this book with the exception that a few impairments to speech are caused by an interruption of airflow right at the level of the vocal cords. Completely interrupting the production of sound ends up impairing speech since without sound, movements of the tongue and lips are not heard. Speech is completely dependent on having a voice so with no voice, there is no speech.

On the other hand you can use a voice without speaking. For example, you can sing without word production. A sound can be generated which is entirely musical in quality and carries no significant cognitive information even though it may carry a great deal of emotional information. A baby's cry is mostly pure sound, carrying some very basic information – and most of that information is emotional and the meaning must be inferred.

We can imagine speech (at an unvarying pitch and volume) as a package of cognitive information. Voice is primarily used for the transmission of speech and adding emotional information. Voice then typically carries information from one location to another. With voice, we increase volume or change pitch to carry our words out to a distance or through background noise. We also alter our volume and pitch to transmit emotion.

Mr. Stillear makes sound, and makes it loudly enough. He just cannot transmit clear content because of his tongue and lip weakness. He does not vary his pitch or volume very much and he comes across as having not much emotion. He can be heard, but not understood. He has a problem that the physician will find above the Speech Line. Problems above the Speech Line are not voice problems.

Hoarseness

We all know it when we hear it

Faith N. Metsan showed up at Dr. Goodew's office saying she was hoarse. Like Faith, most of us have a general idea about hoarseness, though perhaps without really knowing what it is in any detail. A performer often knows when her voice is not working well because the change is sudden. She has thousands of hours hearing her own voice for comparison. Yet, singers and performers stake their entire careers on their voice and many have never seen nor had any concrete visual concept of what was transpiring inside their necks.

Yet even for a singer, a very slow change is not very noticeable. She would tend to gradually alter her vocal closure during the slow development of a voice problem, effectively compensating for or hiding the change. She will often attribute vocal problems to a change in technique. Quite often, someone else recognizes the hoarseness first, because it is a change from the last time he heard her sing.

Hoarseness is a common complaint at the doctor's office. Where is the hoarseness coming from? What specifically is wrong when we are hoarse? Without a definition of hoarseness, we don't really have an answer to any of these questions.

In a pure science mode, I could technically view hoarseness as an impairment of the signal-to-noise ratio of the human voice, yet a more descriptive definition might be more helpful in the clinic. Consequently, I think of hoarseness as an *unwanted leak of air though the vocal cords or the irregular leak of air through the vocal cords*. This is an often neglected, but incredibly helpful definition.

Again, there are only two ways to be hoarse:

1. Air is leaking when you don't want it to leak.

2. The vocal cords are vibrating irregularly.

That is it. That is the essence and foundation of this entire book. In every patient who complains of hoarseness, we will find one of these two problems: air leak or irregular vibration or some combination of both. Fortunately, it is possible to see air leaking or vocal cords vibrating irregularly, at least with the technology available today.

For Faith though, until someone puts a mirror or a camera into her throat, the diagnosis about what is making her hoarse is probably a guess, perhaps a simple guess, a statistical guess or perhaps an educated guess. However, even without an endoscope, let's try to improve on this guess since anyone, including you, can listen to the voice. We can hear air leak and we can hear irregular vibrations.

First, let's see how a human voice is produced before we elaborate on this definition of hoarseness, even though we all know a frog when we hear it.

Voice without vocal cords

Voice is a vibration within the audible frequency range (roughly 60 to 1,100 vibrations per second [or Hz]). In the human, voice is largely produced by the true vocal cords, but it is possible to vibrate other structures and create a voice.

I have a set of wind chimes hanging from an arbor that catch my attention whenever I am out in the garden and the breeze kicks up. They were given to me by Mrs. Mary Marlboro's niece. Mary had purchased them while in hospice with instructions for her niece to give them to me after she passed on. I had cared for Mary for several years after I removed her larynx because of a cancer.

Throughout her life she loved to talk and when growing up had associated social conversation with smoking. The years of tobacco smoke moving over her vocal cords caught up with her, encouraging a few cells to grow without stopping and by the time I met her with a hoarse voice, there was a fairly large lump growing from her vocal cord. Although I removed the growth, then treated the remainder of her larynx with radiation, her cancer persisted and ultimately, I completely removed her larynx trying to cure her of cancer.

Initially after removal of her vocal cords, she used an electrolarynx, an electronic communication device. When held against her neck, the vibrations generated by the device resonated inside her throat creating a voice which she could use to produce speech. The device produced a single tone at a single volume. Her speech was understandable, however, she had a robotic sounding, electronic voice.

Still, she loved to talk. Over the course of a year she worked hard to develop esophageal phonation, learning how to swallow air and breing it back out of her stomach at will. The soft and flexible esophagus (swallowing tube) vibrates slowly generating a deep pitched sound. She could belch fast enough to carry on quite a conversation that was a little less monotonal than the electrolarynx. The desire to talk seems to be hard wired into some individual's brains. Mary found a way to talk no matter how big the obstacle seemed.

The problem for Mary was that the electrolarynx vibrated at a single pitch and her esophagus vibrated at only a few, very low pitches. She could move her tongue and palate and produce words, but she sounded robotic or uninterested. Her electronic or esophageal voices were unable to convey much emotion, even though the content was there. She was missing the use of her original and typical vocal range.

Voice, then, can be thought of as the underlying signal on which the speech or "information" is carried. The signal has two predominant characteristics that can be altered by the larynx. They are pitch and volume. The vocal cords are particularly talented at altering these characteristics which in turn affects how far our sound will travel and the emotion that will be carried with it.

The vocal cords are quite good at putting out a strong signal if desired. A well-produced sound can carry information quite a distance. I can think of a baby in a church, my daughter screaming or an opera singer still heard clearly in the cheap, upper balcony seats.

The vocal cords are quite talented at demonstrating emotion. Think about sighs, whines, giggles, laughs, growls and all the other sounds we make that are not words but strongly convey emotion.

I can be digging in the garden when a small gust sets the wind chimes ringing. When I hear the chimes, Mary is still talking to me. I am reminded of how much she loved to talk. The five pipes are tuned to different pitches, calling out to me with a touch of emotion, reminding me of how, when Mary lost her vocal cords, she recovered her ability to speak, but she still lost the ability to communicate her emotions easily.

Anatomy of the voice box

Sipping espresso in Ken's Boulangerie & Cafe, I hear the background noise of the espresso grinder mingling with the murmur of voices. The couple next to me are chatting in Japanese, drinking coffee and downing a pastry. I don't understand a word, but I can hear that each person's voice box, or more technically the larynx, is functioning well.

The larynx is designed to control the flow of food, air and sound. Each person takes a sip of coffee, swallows, breathes in and then engages in repartee. Liquids and solids are separated from the air each individual breathes, then directed down the esophagus. Air is directed between the vocal cords into his and her lungs. The air is put to use again to make sound on its way back out. The larynx does all of this work, all very rapidly.

Thus, the larynx is basically a valve, a talented valve, with three functions:

1. regulate breathing,
2. create sound and
3. keep food and liquids out of the lungs.

Of course, we have all seen someone blast his friend with drink when he rushes his larynx and tries to swallow and talk simultaneously. It helps to have a microsecond or two between functions.

As the young man beside me swallows his coffee, I see a bulge moving up and down on the front of his neck. We call this protuberance in a man's neck, the "Adam's apple" or medically speaking, the thyroid cartilage. The thyroid cartilage starts out soft and enlarges when exposed to testosterone, perhaps partly for evolutionary mating reasons. The net external effect is that it protrudes visibly. On the inside testosterone thickens and elongates the vocal cords. The longer and thicker they end up, the lower the notes they are capable of producing. The thyroid cartilage sits on top of the cricoid cartilage. Below the larynx is the windpipe or trachea which you can feel in some people with thin necks. Above the Adam's Apple is the hyoid bone which helps suspend the larynx in the neck.

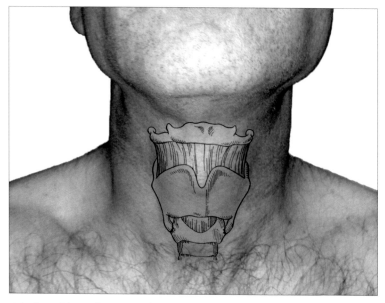

The hyoid bone (top yellow) is connected to the tongue and jaw muscles above and the thyroid cartilage below. The thyroid cartilage (central blue) is also known as the Adams Apple and protects the vocal cords which are attached on the backside in the middle. The cricoid cartilage (lower green) is a complete ring supporting the bottom of the larynx.

In this book, our views of the larynx will be internal from the point of view of an endoscope, essentially the view from above, which is available for easy access by anyone with an endoscope. A rigid endoscope views the larynx from the back of the mouth. A flexible endoscope views the larynx from the back of the nose.

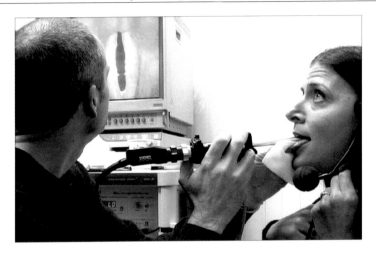

A rigid endoscope is passed through the mouth to visualize the vocal cords. It is difficult to say much of anything other than "eeeeee" while the examiner holds your tongue, but it does provide a very clear picture of the vocal cords. Most people need some topical anesthesia in order to keep from gagging during this examination.

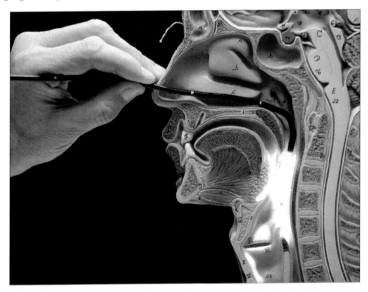

A flexible endoscope is passed through the nose. At the back of the nose, it is angled downward to view the vocal cords.

The space inside the throat above the larynx is the pharynx. It is surrounded by muscles and the hyoid bone. Changing the shape and size of the pharynx alters resonance.

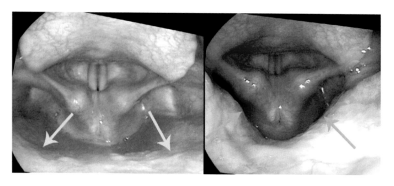

Left: *The pharynx is the opening around the larynx and here it is relaxed and open (yellow arrows).* **Right**: *The pharynx is being squeezed to amplify or improve the resonance of high pitched sounds.*

Technically the larynx has two cartilages that maintain its shape (thyroid and cricoid). A softer cartilage acts as a diverting valve during swallowing (epiglottis). Two smaller cartilages open and close the vocal cords (arytenoids) and some miniscule cartilages sit on the arytenoids seeming to act as a dam to prevent residual liquids in the throat from entering the airway.

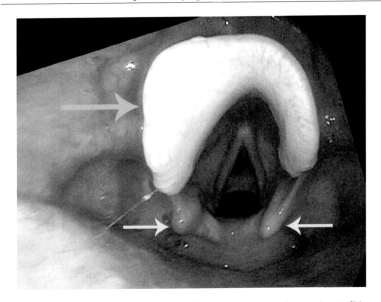

The soft curved cartilage in the middle of the photo is the epiglottis (blue arrow). It's base is attached to the back of the thyroid cartilage just above the vocal cords (the inverted V in the middle of the picture). The epiglottis comes in many different shapes and it folds over when swallowing to cover the vocal cords. The corniculate cartilages rest on top of the arytenoid cartilage (yellow arrows).

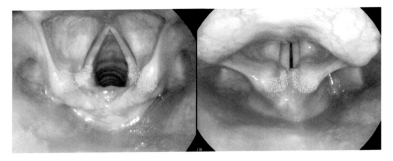

Left: *The arytenoids (approximate location colored yellow) open. The round tracheal rings are partially visible beneath the vocal cords.* **Right:** *The arytenoids bringing the vocal cords together to make sound.*

The front of the thyroid cartilage is triangular or tent-like in shape suspending and protecting the vocal cords, with the Adam's apple representing the apex. Inside, the airway is essentially a round tube with the vocal cords narrowing the airway, acting as a valve, technically – the glottis. The vocal cords narrow the opening to a triangle. During exhaling (breathing out), the vocal cords narrow the triangle to keep some back pressure in the lungs. During phonation, the vocal cords come almost completely together to form a narrow slit. Air passing between the vocal cords sets them vibrating and generates sound.

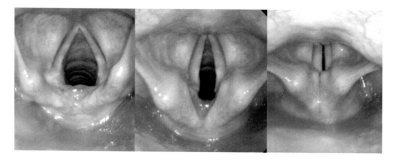

*Left: The vocal cords form a triangle when breathing in, **Middle:** narrow the triangle when breathing out and, **Right:** line up parallel to form a slit during sound production.*

There are 10 internal muscles, vocal ligaments and some glands for lubrication. This is all covered with delicate, nearly translucent mucosa, tinted pink when viewed from far away by the blood flowing beneath it. All of this sounds complicated, but the machinery of the voice box is elegantly simple, with each component serving a specific purpose. Let's take a closer look at the muscles, or if they are deep beneath the surface, we will visualize their action by their effects in the photos.

Intrinsic muscles of the larynx

The larynx is comprised of 10 muscles. With the various muscles changing the position, length and tension of the vocal cords quite a range of sounds can be generated.

There are five muscles on each side of the larynx.
TA – ThyroArytenoid
LCA – Lateral CricoArytenoid
PCA – Posterior CricoArytenoid
CT – CricoThyroid
IA – InterArytenoid

Since each muscle is paired, any asymmetric contraction represents a probable weakness. Understanding the function of each muscle will help one understand neurologic and muscular injuries especially well.

First though, we can think about the general operation of the vocal cords. During phonation at a low pitch, the vocal cords are brought together, but tension remains relatively loose leaving the vocal cords short and thick. At a high pitches, they are conversely tight, long and thin.

The muscles are hidden from our direct view, though in some cases we can see the bulk of the muscle beneath the surface. Additionally, since each muscle can do only one thing – contract – we can visualize some effect from each muscle during endoscopy if we know how to trigger contractions of that muscle.

One complicating factor is that the larynx has some redundancy in this setup and when a person develops a voice problem, she naturally compensates in the most expeditious way possible to maintain her voice. For a laryngologist, testing the individual muscles also involves unmasking this compensation.

ThyroArytenoid muscle (TA)

The TA muscle lies within and runs the length of the vocal cord. The muscle provides most of the filling or mass of the vocal cord. It tightens to increase the pitch, mostly by isometrically tensioning the vocal cord. For singers when they are not engaging the CT muscles, the TA muscle essentially is used to change pitch throughout their lower or chest register.

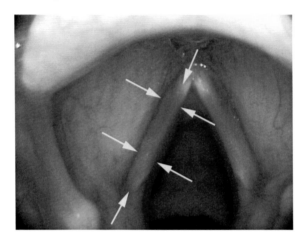

The thyroarytenoid (TA) muscle creates the bulk of the mass within the vocal cord. Arrows outline the generally visible mass of the muscle when viewed from above.

We can see the bulk of this muscle during endoscopy on a healthy vocal cord. The easiest way to appreciate the bulk is in the patient where one vocal cord is completely paralyzed, especially when viewed along the length of the vocal cord. In a complete paralysis, the paralyzed vocal cord would consist of only the mass of the vocal ligament. The difference in size between these vocal cords then represents the mass of the thyroarytenoid muscle.

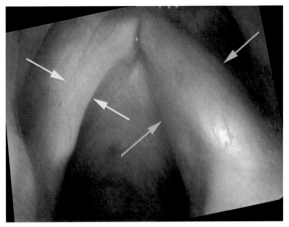

A view from nearly behind and nearly parallel to the vocal cords, looking along their length (almost a cross-sectional view). The left thyroarytenoid muscle is atrophied (yellow arrows) while the right muscle is a normal size (blue arrows) demonstrating how much of the vocal cord bulk consists of the thyroarytenoid muscle.

CricoThyroid muscle (CT)

The CT muscle is located on the exterior of the larynx, just under the skin. The thyroid cartilage pivots or rocks forward on top of the cricoid cartilage when the CT muscle contracts.

Below the thyroid cartilage and above the cricoid cartilage is a groove. Before the Heimlich maneuver became popular, many people knew of this groove as the location for placing a hole in the neck in the event someone was choking on food at dinner time. There is not much tissue between the skin and the airway here and it is actually a very safe place to make a hole for breathing in an emergency.

If you place your finger in the groove and attempt a high pitch sound, you will feel this groove pinch closed as the cartilages are pulled together. The CT muscle pulls these cartilages toward each other and is located on either side of the middle groove.

The effect of this rocking is to stretch the vocal cords. For a singer, the additional tension provided by stretching the vocal cords provides the falsetto (upper or head) register. The tension of the CT muscle also allows us to yell with less effort.

Some common complaints when this muscle is not working are:

- inability to sing as high as in the past,

- additional effort to project as loud,

- discomfort with projected speaking.

This muscle is not visible directly on endoscopy since it is outside the larynx. The effect of the muscle is easily visualized as a lengthening of the vocal cord as the pitch is increased during phonation.

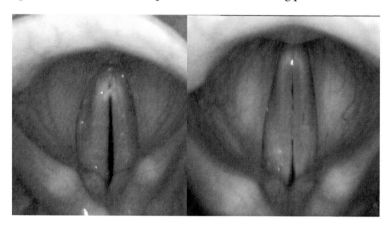

Left: the vocal cords are vibrating at a low pitch. Right: at a high pitch the CT muscle has stretched the vocal cords.

The fact that we have two muscles to change pitch, one primarily low in the range (TA) and the other primarily high in the range (CT) means that there is a mid-range location where the tensioning of the two muscles blend together. For singers, this is the area of the vocal break. Classical singers spend a great deal of time smoothing the transition between the use of these two muscles.

Lateral CricoArytenoid muscle (LCA)

The LCA muscle brings the vocal cords from an open breathing position together to a closed position in preparation for making sound. This muscle is located parallel and just lateral to the TA muscle within the vocal cord. It is not particularly visible on an endoscopic exam although the effect of the contraction of the muscle is easily visualized. The LCA muscle is attached to the outer end of the arytenoid cartilage which acts like a lever. The other end of the arytenoid lever is the vocal process, a white cartilage visible towards the back of the vocal cord. Contraction of the LCA muscle rotates the vocal process to the midline and if necessary beyond the midline.

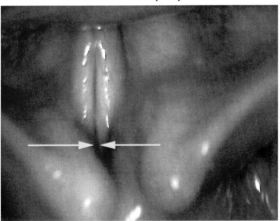

The LCA muscles have brought the vocal cords completely together.

When relaxed, with the other muscles functioning normally, the vocal process is at the apex of an approximately 160°–angled, gently curving corner, formed by the membranous vocal cord and the soft tissue posterior to it (see photo in PCA section below).

During normal contraction this corner disappears to form a straight line although most of the time the vocal processes are covered up by the arytenoid itself with the vocal cords closed. The normal configuration of the vocal process during LCA contraction is best understood by viewing the configuration when the LCA muscle on one side is injured and that vocal process cannot move to the midline. Compensation may occur. The opposite healthy LCA muscle has or can develop enough strength to move the vocal process across the midline and many times can reach or approach the vocal process on the weaker side. This may result in an inversion of the angle such that the vocal process now protrudes into the airway.

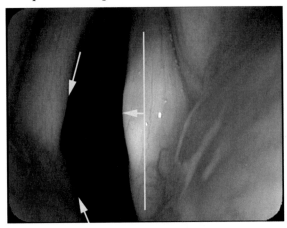

This photo is an ultra close-up view of the vocal processes at the back of the vocal cords taken while this patient with a left LCA muscle paralysis is trying to make a sound. Because of weakness on the left, the right vocal process is compensating by pushing (short arrow) beyond the midline (vertical line), while the left vocal process remains canted laterally, forming an obtuse angle (left arrows). If this patient's vocal cords were healthy, they would be lined up parallel, along either side of the central line as in the previous photos.

InterArytenoid muscle (IA)

The IA muscle holds the vocal cords in a closed position after they have been brought together by the LCA muscle for phonation. The IA muscle from one side attaches with the IA muscle from the other side. Because the IA muscles connect in the midline, this is one muscle where there seems to be cross innervation or nerve input from the opposite side. So even when there is complete paralysis of one side of the larynx, the IA muscle on the injured side may still function because of this cross innervation. In this situation, on endoscopy we may see a twitch of the arytenoid of an otherwise apparently completely paralyzed vocal cord due to innervation from the opposite IA muscle.

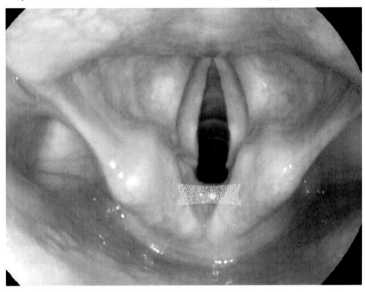

The interarytenoid muscle (yellow bar) is located between the arytenoid cartilages in the midline at the back of the glottis.

Posterior CricoArytenoid muscle (PCA)

The PCA muscle is located on the back of the larynx, behind and lateral to the arytenoid cartilage. It pulls on the arytenoid cartilage to open the vocal cords for breathing. The effect of contraction of this muscle can be visualized as the exact opposite effect of contraction of the LCA muscle. The PCA muscle is strongly activated by sniffing. During a brisk inhale through the nose, the PCA muscle contracts and each vocal process moves laterally, increasing the size of the opening of the glottis.

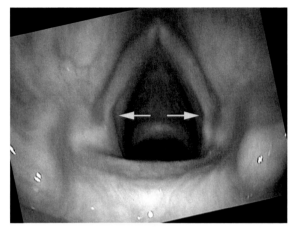

During a brisk breath in, particularly through the nose, the PCA muscles contract and the vocal processes (arrows) move far laterally, opening the airway to its maximum size. This is the configuration when the LCA muscles are at rest.

Additionally contraction of the PCA muscle can be viewed when a person initiates sound with excess tension. The LCA muscle brings the vocal cords together but then the PCA muscle also contracts and pulls the vocal processes slightly apart. The actual belly of the muscle can be seen bulging behind or posterior to the arytenoids.

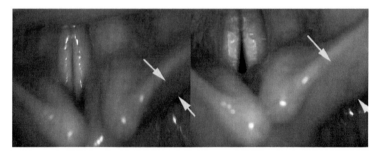

Left: *the relaxed right PCA muscle's bulk between the left arrows is small.* ***Right***: *the right PCA muscle has contracted, (pulling the vocal cords slightly apart) and the increased thickness of the right PCA muscle is visible between the right arrows.*

Compensation

During an examination people hate to be embarrassed and naturally and fairly immediately try to compensate for any hoarseness. It is fortunate that there is redundancy in the vocal system so compensation is usually available to maintain sound production even when there is a problem. However, for the examiner, compensation tends to hide a voice disorder, both audibly as well as visually. Some common instances hidden by compensation include mucosal disorders on the edge of the vocal cord or weakness from neurologic and muscular disorders. Yet the astute examiner eliminates compensation to visually expose a problem.

For example, there are 2 muscles, the TA muscle and the CT muscle, which both have the effect of raising pitch. The TA muscle is within the vocal cord and isometrically tightens the vocal cord to raise pitch, while the CT muscle is located on the external surface of the laryngeal cartilages and lengthens the vocal cords to raise the pitch. If there is decreased neurologic input to one of these muscles, the other muscle compensates at least partially. In the case where the TA muscle is not working on one side, a patient automatically compensates by pulling the CT muscle tighter. This has the effect of raising the speaking pitch when the patient tries to speak with the same volume as before the injury.

When I perform an endoscopic exam on a patient speaking at an unnaturally high pitch, I really want to hear the patient produce a lower and lower pitch. At each lower pitch, compensatory tightness from the CT muscle is progressively removed. As the CT muscle's contraction is removed, if there is a weak TA muscle, the vocal cord would begin to bow and then flutter in the wind while a healthy TA muscle on the other side would remain tight. Removing compensation amplifies this difference between each side of the larynx and allows improved visualization of this neurologic impairment.

Mucosal disorders – problems with a swelling on the edge of the vocal cord – also involve natural compensation. At higher pitches, the vocal cord is pulled tighter. A mucosal swelling on the edge of a vocal cord would stand out farther and farther with increasing pitch. The further it sticks out, the more likely the swelling is to touch the other vocal cord and stop vibrations. The patient compensates for this "stopping effect" in some instances by increasing the airflow from below to blow the touching swellings apart. In effect, the patient increases their volume to compensate for effects from the vocal swelling sticking out along the edge of the vocal cord. A singer with a swelling will sing louder and louder as she goes up in pitch to prevent the vocal stoppages. As an examiner, I remove this compensation by asking the patient to make only soft sounds as she goes up in pitch. At low lung pressures, the slightest touch of a swelling on one vocal cord stops or alters the vibrations on both. I remove the volume compensation in order to better hear, visualize and discover the pathology.

There are other forms of compensation which will be discussed in some of the case studies in later chapters. Keys for the examiner:

- we automatically try to compensate and maintain normal vibration as much as possible,

- removing compensation, – typically by changing pitch or volume – asymmetries and gaps become more audible and more visible.

Vocal cords, vocal folds

What we term "vocal cord" is often compared to the edge of a string when viewed from above during vibration and that is likely where the term came *cord* came from. I also frequently see the term "vocal chord" though I believe *chord* is a misspelling as the vibrating portion of the larynx does not generally represent a portion of a curve nor three musical notes. Viewed in cross section the vocal cord doesn't look like a cord at all. It is more of a wedge in shape. Still, musically it does function like a cord and many analogies to a cord or string, such as a comparison to the string on a guitar, can be made.

Some people call them "vocal folds," which is a more apt description of their three-dimensional visual appearance. They are a fold of soft tissue rising from the edges of the airway. However, the terms vocal cord and vocal fold can be, and are, used interchangeably.

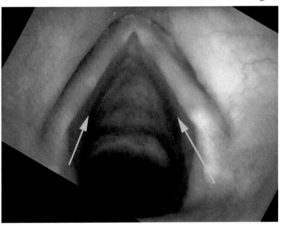

A fullness (arrows) can be seen below the edges of the vocal cords that actually gives the vocal cord the shape of a fold along the wall of the airway. The fullness is from the thyroarytenoid muscle.

The vocal cord is essentially a muscle under a ligament with mucosa covering both of them. A layer of lubricant, the lamina propria, lies between the mucosa and the ligament-muscle combination. The muscle tightens and loosens to change the pitch of the vibrations. The lubricating layer allows the mucosa on the surface to vibrate easily. The muscle oscillates to a small degree, like a string on a guitar or piano, but the lining on the surface is the primary oscillator or generator of sound. The mucosa is like a layer of silk draped over the edge of the ligament and the mucosa vibrates when air passes rapidly by it.

When we examine sound production, we will be able to establish typical configurations of the voice box for each of these problems. The vocal cords have two positions: ABducted and ADducted. When ABducted, they are open for breathing and in the configuration of an inverted V.

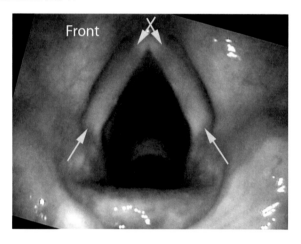

Here the vocal cords are viewed from above in the ABducted, "breathing in" position. The vocal cords are the white bands on either side. The V at the front (anterior) is attached to the inside of the thyroid cartilage or Adam's apple. The arrows point to the left cord, which is on the left side of this picture. The thyroarytenoid muscle within the vocal cord is attached to a slightly visible white ligament at the front and to the tip of a white cartilage at the back (the vocal process of the arytenoid cartilage).

When ADducted the vocal cords should come essentially into alignment parallel to each other. The lungs generate pressure below and air can then be passed through the vocal cords from the windpipe below. As air passes through this narrow slot between the vocal cords, the mucosa starts to vibrate, creating sound.

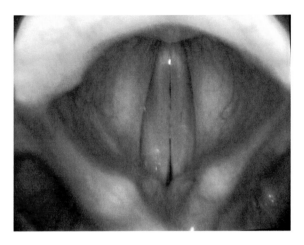

The vocal cords are the two white structures lying parallel to each other in the center of this photo taken during sound production. The vocal cords are in the ADducted position.

Let's define how the vocal cords produce sound when all is normal.

Vocal cord vibration

Voice is sound. Sound production is based on physics. All vocal impairments occur because of a physical change in vibration.

In the idealized situation, sound is made when:

1. the back of the vocal cords leave the "breathing in" position, moving together until parallel,
2. tension is applied to the vocal cords as they initially occlude the airway,
3. air is propelled through them, usually from below,
4. they start flexing open in the middle and
5. increasing tension causes the cords to snap back closed. The last two steps repeat over and over, creating pulses of air – vibration.

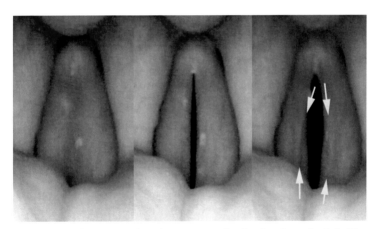

Vocal cords vibratory cycle. They are completely closed on the left. They begin to open in the center and on the right they have reached maximal opening before they will begin to close again. The mucosal wave can be seen in the right photo as a curved line just lateral to the edge of the vocal cord (arrows). In a video, this line propagates from the margin and moves laterally.

The vocal cords oscillate quite rapidly, perhaps 100 to 200 times per second during casual speaking, with smaller cords tending toward faster oscillation. In a set of perfect cords, we could characterize them as:

- being open about half the time and closed about half the time,
- letting air out in measured puffs,
- not leaking air during the closed phase and
- vibrating regularly
- vibrating symmetrically.

This creates the sound that we hear and any single note can be visualized on an oscilloscope as a sine wave – a regular vibration and when we hear it, we hear a musical tone. We can talk about the tone in terms of frequency, often measured in Hertz or vibrations per second. Hertz is a common scientific measurement that requires the use of logarithms for calculations.

We can also use a musical scale such as the chromatic scale, composed of 12 equally spaced pitches, to label each tone produced (C3, C3#, D3, D3#, etc)[3]. Each succeeding note is one semi-tone higher

---

3   Measuring pitch and pitch notation.

When I evaluate a patients pitch by ear, I match the pitch to a note on a piano. Middle C on the piano is C4 – that is the fourth octave on the piano. One octave lower is C3. The numbering changes at C so the notes in an octave would be labeled

C3 C3# D3 D3# E3 F3 F3# G3 G3# A3 A3# B3 C4 C4#...

Though octave means eight and there are eight steps in the Western diatonic scale in music, there are 12 "half-steps" in this chromatic measuring system before returning to the same note (C4 has double the number of vibrations as C3). We describe these as "equal intervals". To our ear, the distance between C3 and C3# is the same melodic interval as the distance between C4 and C4#, as well as the distance between any other half-step.

However, if one uses the Hertz scale for measurement, the distance of one semitone between C3 (130.81 Hz) and C3# (138.59 Hz) is 7.78 Hz. The distance of one semitone between C4 (261.63 Hz) and C4# (277.18 Hz) is 15.55 Hz. The Hertz scale is a nonlinear, logarithmic scale and not easily added and subtracted. Arithmetic manipulation can be done, but the simplicity of using "half-steps" far outweighs the precision of the Hertz scale for clinical diagnosis.

than the previous. I like this "semi-tone method" for documenting the voice since the visual distribution of keys on the piano separates the sounds into audibly equal intervals without delving into the complexities of logarithms.

Singing

Singing in a western classical music style, particularly in opera, the singer's voice quite often approaches a very pure, clear musical tone. With the vocal cords roughly capable of vibrating between 60 to 1,100 times per second when singing, we hear a wide range of sound. At times, particularly at high pitches, the voice has only a musical quality and the speech aspect cannot be easily discerned. Think of the diva soprano singing an aria at the top of her range. This style of pure tone singing requires the most regular, most precise positioning and closure of the vocal cords.

By contrast there are popular or folk musicians who may have a very soft quality to their voice. Think of the sexy country or steamy cabaret singer. These particular types of musicians typically hold the back of their vocal cords apart so that air leak is part of the style.

There may be extreme air leak or even irregular vibrations in some genres. The false vocal cords can also be enlisted to vibrate and produce sound. Tuvan throat music or Heavy metal screaming may use these qualities to attract listeners.

Singing combines voice with speech to varying degrees depending on the style. A rapper could conceivably perform an entire song on a single note, emphasizing the spoken aspect of the music.

In singing, the emphasis on how the vocal cords vibrate generally rises above the words alone. In singing, we wish to do more than just communicate information. Varying the pitch and the volume adds a great deal of emotion. The singer intentionally varies sound production to produce these various tones and different degrees of air leak. These various forms of sound production deviate from the pure opening and closing described in the previous chapter, and when utilized intentionally are not impairments, they are music.

Two types of hoarseness

The key to diagnosis

Air leak & diplophonia

All hoarseness can be described in one of two ways. In the first, the vocal cords do not come together as they should, allowing air leak. Let's call this *husky hoarseness*. The second type of hoarseness is caused by asymmetric vibration. Because there are two vocal cords, when they are not symmetric, they tend to vibrate at two different pitches. A physician would call this sound diplophonia. *Diplo-* meaning double and *-phonia* meaning voice, so two voices can be heard at once. Usually, since the cords are only slightly out of sync, the diplophonia will be inharmonious and at our typical, rather low speaking pitch, the perception will be of a rough or gravelly quality. We will call this *rough hoarseness*.

Let's explore these two ideas

- **husky hoarseness and**
- **rough hoarseness**

in more detail.

Whisper

Pure air leak

To understand husky hoarseness, it is helpful to think of one of the extreme types of sound production we can make with our larynx: extremely soft sound. We can generate sound in the larynx with the vocal cords in a partially or completely open position – that sound is called a whisper.

Normally the vocal cords come completely together when making a sound. The mid-portion oscillates open and closed.

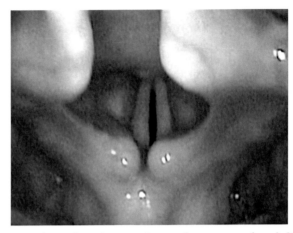

During normal sound production, the membranous vocal cords lie parallel to each other and vibrate – opening and closing. Although the back of the vocal cords are hidden from view beneath the arytenoids in the photo, the back of the vocal cords are together. This stroboscopic photo was taken during the open phase of vibration.

By contrast, in a whisper, the vocal cords do not vibrate. The airflow is increased and as it passes a narrowed spot, turbulence is created. Airflow that is turbulent consists of many different pitches simultaneously: white noise. The larynx can narrow the airway in several ways without allowing the vocal cords to vibrate.

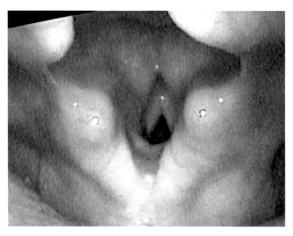

In a whisper, the back of the larynx might be open while the membranous vocal cords might be closed such that all the air is forced out through an opening at the posterior end of the vocal cords.

This signal, composed of white noise, lacks intensity and cannot be heard well or from very far away. It is not very penetrating. White noise blends with environmental noise. For example, open the car windows while traveling and a great deal of turbulent airflow is created at the window. This white noise effectively masks other sounds. A whisper, which is also white noise, in a car blends in with the sound created by the open windows and cannot be heard at all.

Conversely, we typically produce voice at a single pitch to generate a strong signal that will stand out against background noise. This is a signal that we modify with vowels and consonants to carry information from one human to the next. With a clear tone we can generate sounds that are distinct and carry well, even in a convertible on the highway driving with the top down.

A stage whisper is meant to sound like a whisper, yet the audience needs to hear the speaker. In a stage whisper for a theatre production, the vocal cords are allowed to vibrate a little, so that the sound has the character of white noise, but enough vibrations of the vocal cords for the sound to carry into the audience. A stage whisper is really a mixture of a lot of whisper and a little bit of a vibration.

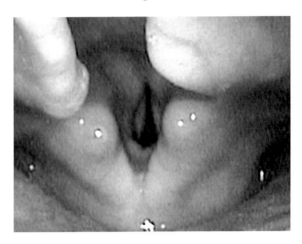

In a stage whisper, the membranous vocal cords are held slightly apart such that much of the air exits through the opening behind the vocal cords. However, some air passes between the vocal cords and they vibrate slightly.

In between a whisper (cords not vibrating) and a clear tone (cords completely parallel with all air passing between them), we can have some mix of a pure tone and white noise. The white noise gives the voice a husky quality.

Later in this book, we will examine individuals where this huskiness is not desired and air is leaking out. Other times, for example, a nightclub singer might wish to add a component of breathiness to her voice to give it a sexy quality. She is adding an intentional gap between the vocal cords to let some air leak out. A novice singer may be straining so hard, she ends up tensing multiple muscles in the larynx and inadvertently holds the vocal cords apart.

So while a whisper is the extreme of turbulent airflow through the larynx, any gap between the vocal cords will add a commensurate degree of turbulence perceived as a huskiness of the voice. We may desire that quality, but if we don't, it is hoarseness.

Gaps

Mind the gap

Getting around the city of London on the Tube also known as the London Underground, each time the train stops and the doors slide open, both in and outside the train an announcer comes on with the slogan, "Please, mind the gap." Signs on the walls remind passengers to "Mind the gap." Over and over one hears, "Mind the gap." The gap between the train and the platform can be the source of all sorts of serious problems, but is otherwise ignored by the millions riding the rails around London.

A laryngologist would do well to ride the tube until, "Mind the gap" becomes second nature. The gap between the vocal cords deserves your primary attention, and it is easily forgotten looking at all the other structures around, although the gap is the most significant location to find the source of hoarseness. So, mind the gap!

Remember, under idealized conditions the posterior portion of the vocal cords move completely together during voice production and the vocal cords lie parallel to each other. Air is blown between them and the vocal cords rapidly open and then completely close, letting out regular pulses of air.

In what ways might the vocal cords incompletely close? There are a couple of characteristic types of gaps that occur between the vocal cords:

- Posterior gap,

- Anterior gap,

- Central gap,

- Split gap and

- Timing gap.

All of these types of gaps leak air and create a husky hoarseness.

Posterior gap

If the vocal cords are going from the breathing in, V-shape toward a parallel position but they stop short of complete closure, this leaves a gap posteriorly. One example of this type of gap is muscle tension, where the opening muscles (PCA) partially tighten during phonation and hold the cords slightly apart, allowing air to escape between the back portion of the vocal cords.

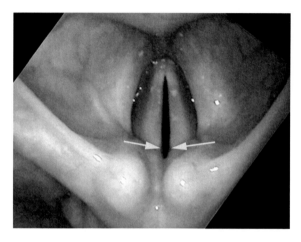

Arrows lie along the direction of closure and point to the gap remaining between the vocal processes when there is incomplete closure.

Anterior gap

Vocal cord trauma may disrupt the vocal cords where they attach at the front of the larynx. If the vocal cords heal slightly apart anteriorly, air escapes through the front of the vocal cords. It really becomes impossible for a person to close this gap without the help of surgery.

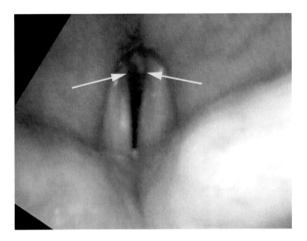

Arrows point to an anterior opening, secondary to scarring after a surgery on the larynx that injured the attachment of the vocal cords to the thyroid cartilage.

Central gap

If the vocal cord muscles are not exercised regularly by talking (imagine a lighthouse keeper sitting alone on an island year after year), the vocal cord muscles atrophy and can no longer tense to a straight line. They remain concave, even when the back of the cords are completely closed. Consequently, a central gap is created.

Aging contributes to this as well. Much like the skin on the face gradually sags with aging, the vocal cords sag with aging as they lose elasticity. Typically this sagging, or as physicians call it, bowing, is relatively symmetric. There is a nearly oval shaped gap with pointed ends between the vocal cords. Air leaks out the middle.

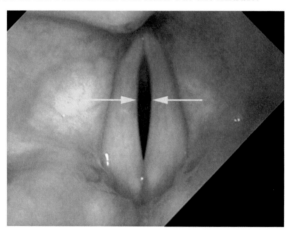

At the beginning of sound production, the back of the vocal cords have touched, but the arrows point to a large central gap.

An asymmetric central gap is created with a unilateral (one-sided) recurrent laryngeal nerve injury. (For further discussion about the nerves supplying the vocal cords, see "Neurolaryngology" on page 181). The vocal cord muscle on the side of the injured vocal cord atrophies and is not as tight nor thick as the other side, so at any given attempted pitch there is air leak in the central portion of the vocal cord, albeit with more leak on the paralyzed side.

Split gap

Vocal overdoers typically have hefty vocal cord muscles and may also have a callus or swelling on the edge of the vibrating portion of the vocal cord. This swelling is typically in the very center of the cord and it stands out from the edge of the vocal cord. This protuberant swelling (or swellings) will touch first as the vocal cords are tensed and leave an opening both anterior and posterior to the swelling. Hence air will leak from in front of and behind the swellings.

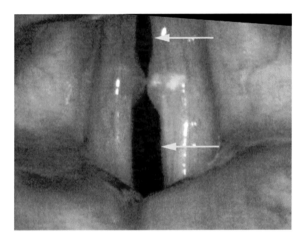

Two vocal polyps, slightly different in size, touch before the vocal cords can completely close, allowing air to leak from in front of and behind them (arrows).

Timing gap

Anything that makes the vocal cords uneven, such as in mass or tension, can put them out of sync. With a mild asymmetry, they may oscillate out of phase. Under a strobe light, it will look like they are chasing each other. They may never touch each other and so effectively, even while they are crossing each other's path in the midline, they do so at separate times, so air continuously leaks out. This is a gap created by timing.

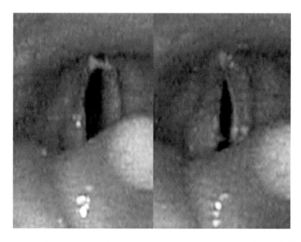

The same vocal cords viewed at two moments in time. In the left photo, the right vocal cord is near the midline and the left vocal cord is at its apogee laterally. In the right photo, the left cord has now come to the midline, but the right cord is now out in a lateral position so there is always an opening.

If they become slightly more asymmetric, they may begin to oscillate at different frequencies. Two pitches are created simultaneously – diplophonia. The separate frequencies are perceived as roughness. And in this case we would actually hear both huskiness and roughness. The huskiness is from the gap. The concept of roughness brings up the other type of hoarseness: roughness due to vocal cord asymmetry.

Asymmetric vocal cords

Asymmetric tension

Returning to Faith N. Metsan whom we met earlier, we may recall that she lost her voice suddenly for no apparent reason. When she went to her primary physician, he predicted that she would improve with time. Indeed she did. After a few weeks, some voice returned. Her voice was higher pitched and softer than before, but seemingly her physician was correct, she would get better on her own.

Still, after several more weeks, Faith was unsatisfied with her merely marginal improvement. At a follow up visit with her primary physician, she was referred to an otolaryngologist. Still she failed to improve with various treatments. Several months later she appeared in my office still hoarse.

As she tells me her story, I listen to her voice. Next, I asked her to start at a high note and gradually descend to the lower pitched notes. At the highest notes, her voice is fairly clear and fairly robust. At a lower pitch, approaching the typical speaking pitch for a female, she loses volume and below that pitch, her voice becomes very rough. We repeat the task several times. Her voice fading in volume and quality at low pitches, is very consistent.

When I look at her vocal cords as she descends toward the lowest pitches where her voice becomes rough, I can see one vocal cord become loose, bow, buckle and ultimately flutter in an irregular pattern.

Faith's rough vocal quality is due to an uneven tension between the vocal cords. The uneven tension is partially compensated at high pitches, remaining subtle and hidden. Her comfortable speaking pitch is automatically higher than it used to be as she unconsciously tries to compensate for her voice problem by tightening the vocal cords to avoid the flutter. In her lower range, however, the difference in tension is pronounced and at some point each vocal cord vibrates at a separate pitch.

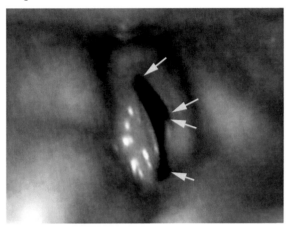

Faith's vocal cords oscillating in a wild pattern because the right side is much looser than the left at a low pitch. The right cord has two separate wave segments compared to the left cord's single wave.

To describe these two separate pitches occurring simultaneously (diplophonia), I could use terms such as roughness or gravelliness to describe her low voice, while she might only say, "I am hoarse" or "My voice is rough at low pitches." Two pitches produced simultaneously and sonically competing with each other are irregular vibrations. Faith's roughness is due to uneven tension.

There are several potential types of unevenness between the vocal cords that may lead to diplophonia:

- asymmetric tension,

- asymmetric mass,

- asymmetric length and

- asymmetric stiffness.

It is also possible to have diplophonia from an irregular pattern of vibration even when the vocal cords are not asymmetric between each other. This is what occurs with vocal fry which produces a regularly-irregular vibration. The vocal folds vibrate at a regular rate and then a beat is skipped, perhaps every third beat, every eighth beat or some other regular interval. This ends up generating two pitches simultaneously.

Let's explore the different types of asymmetries.

Asymmetric mass & length

Norm A. Lee is an unabashed sports fan, with season tickets to several professional leagues in town. He is a vocally enthusiastic supporter at his kids games also. His voice has been rough for several months. He never had any trouble with it until last autumn. In retrospect, he recalls going to a Friday night football game, where he was yelling quite a bit and by the end of the game he had lost his voice. His voice came back to a usable level by Monday and he was able to go to work, but it never regained a normal quality since that event.

His story suggests that yelling and hence vocal trauma might be the likely cause of his voice loss and subsequent hoarseness. When I listen to his voice, he can make the low notes reasonably well, though he says that as he talks more, his voice becomes rougher. Indeed, during the course of our conversation, his voice sounds progressively rougher. When I ask him to glide up in pitch, his voice cracks and squeaks.

Looking at his vocal cords with an endoscope, a large red polyp hangs on one edge of one vocal cord. The polyp adds mass to this vocal cord. This is a hemorrhagic polyp and has blood coursing through it. As he vibrates his vocal cords, the vibrations create a whipping of the polyp back and forth and the polyp fills with more blood as he continues talking, making this vocal cord heavier and heavier. When the mass between the two vocal cords is different enough, they begin to vibrate separately at different pitches. Hoarseness due to an asymmetric mass is most evident at his low pitches.

He has more than one type of dysphonia created by the polyp depending on the pitch he is trying to make, so he provides some insight into the complexities of hoarseness. In the middle of his range, his pitch tends to jump around and he cannot seem to control it. All of us tighten our vocal cords as we move up in pitch. At some specific pitch in the middle of Norm's vocal range, with progressive tightening of his vocal cords, the polyp just barely begins to rub the opposite vocal cord. As it touches the opposite cord, vibrations are thrown off, his voice quits, squeaks (the pitch jumps up) or breaks up into roughness as they are thrown completely out of synchrony.

A further increase in pitch and the polyp compresses against the other vocal cord stopping the vibration in the middle, leaving either end to vibrate separately. If the polyp is precisely in the middle, the segments may be the same length (and less than half the length of his vocal cord). They will each vibrate at the same, very high pitch (typically the jump is about one octave). If the segments are different lengths because the polyp is based slightly off the middle of the cord, we will distinctly hear two different high pitches simultaneously. Effectively he doesn't have a functional middle range. He can make either low or very high pitches.

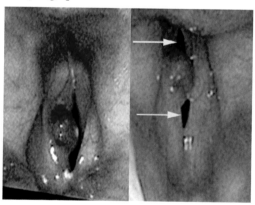

In Norm, at various pitches, we have three types of asymmetries as well as a husky hoarseness:

> 1) *asymmetric mass at low pitches (in the left photo, the left vocal cord is heavier than the right because of the blood in the polyp),*
>
> 2) *asymmetric stiffness (the blood filling the polyp reduces the left vocal cord flexibility),*
>
> 3) *asymmetric segment length at high pitches (in the right photo the front segment is a different length than the rear segment – arrows) and*
>
> 4) *huskiness - air leak on either side of the polyp since the cords cannot come together (the black gaps in the right photo at the arrows).*

In his mid-range, we hear a pitch break, something we normally associate with puberty. The pitch break occurs as the pitch is increased to the point where the polyp just begins to touch the opposite cord. If the vibrations produced by the full cord are suddenly shifted to half the cord, the pitch jumps up. Alternatively, the touching may be followed by a diplophonia if the touching is light enough to just throw the cords out of sync with each other.

Many people quickly adapt to avoid these pitch breaks. By holding the vocal cords farther and farther apart the swelling doesn't touch and disrupt the vibrations. Of course more air leaks and a greater huskiness is created by this compensation (see "Effects of swellings" on page 91) to avoid the pitch breaks.

Asymmetric stiffness

Sarah Trustingham is an amateur singer. She tells me she was diagnosed with a vocal nodule and when she saw a surgeon, he recommended stripping it off. Confident in his abilities, she had vocal cord stripping surgery, but her singing voice never did improve. Since that time, she has just given up singing. However, she would still like to know why she remains hoarse. She asks me, "Have techniques gotten any better? Perhaps you can now fix something that my first doctor could not."

Listening to her speaking voice, it sounds just fine. However, as she starts to produce notes higher up the musical scale, her voice deteriorates and breaks up through much of her middle and upper range. She can only make a few, very clear, somewhat random, high notes.

When I turn on a stroboscopy light and look at her vocal cords in motion, one is very stiff and the other one supple, revealing the explanation for the roughness in the middle of her vocal range. At her lowest notes, the difference in stiffness causes each vocal cord to vibrate at slightly different amplitudes, but still at the same pitch, so no diplophonia. In her mid-range, the vocal cords go out of sync, each producing a different note, so the diplophonia is very audible. At her highest notes the stiff vocal cord is pulled so tight, it does not vibrate at all and all the sound comes from the flexible cord. The tone produced by her single cord vibrating is quite clear.

The stripping of one of her vocal cords appears to have removed not just the mucosa (the surface), but also the lubricating layer of the vocal cord. When she healed after the surgery, it became true that she no longer had an impairment from a nodule sticking out. However, because the lubricating layer was removed by the stripping, the mucosa lining the surface adhered directly to the vocal cord (thyroarytenoid) muscle during healing. Now, her stripped cord is much stiffer than her other vocal cord. Consequently, during many

71

of her attempts to make a sound in her mid-range, the more supple cord is trying to oscillate at a different pitch than the stiffer side. We perceive this asymmetric stiffness as roughness.

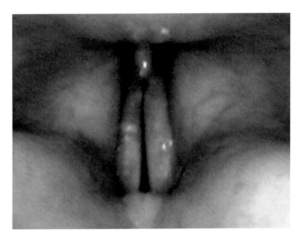

The left cord, remaining straight and stiff, is not moving on a stroboscopic exam, while the supple right cord oscillates back and forth. The left cord also has a slightly darker hue from the scarring.

Asymmetric length

Samantha Wilingsly became very ill, almost dying. When she arrived at the hospital's emergency room late at night, she was no longer breathing and the emergency room doctor intubated her. Intubation means that he inserted a plastic tube down her throat, between her vocal cords, and connected this tube to a respirator to breathe for her while she could not. When she began to recover, the tube was removed, but she again had difficulty breathing and she was re-intubated urgently by the on-call anesthesiologist. She was disconcertingly slow to improve. Each time the breathing tube was removed, she deteriorated, requiring another intubation by another caregiver. When the tube was taken out for the fifth time, she seemed to have recovered her lung function and finally left the hospital.

At home, she found that she wasn't quite back to normal. Several months later she came to my office with a rough and weak voice.

"I run out of breath easily. I cannot exercise like I used to and when I try to talk, I just gasp after awhile. My voice never improved."

With this history, I consider several possibilities. It is possible that on one or more of the tube insertions something was injured in the urgency of putting the breathing tube between the vocal cords. I have seen a torn vocal cord after an intubation. Perhaps as the tube was lying between the cords, some tissue ulcerated and scarred on one side. It is even possible that the *cuff* on the endotracheal tube, which is filled with air to seal the breathing system while on the respirator, may have been inflated too tightly. The cuff, if allowed to sit just beneath the vocal cords, can put pressure on and injure a branch of the nerve operating the vocal cords, in turn causing a paralysis or paresis. So there are a number of plausible reasons to consider as a cause of her vocal impairment.

When I look in her throat with the endoscope, the vocal cords don't completely come together, so when she goes to initiate a sound, this gap explains the softness of her voice. The left cord is also longer than the right, so they do not line up well when they do come together. Her two cords are different lengths which will then tend to vibrate at different pitches explaining the roughness in her voice.

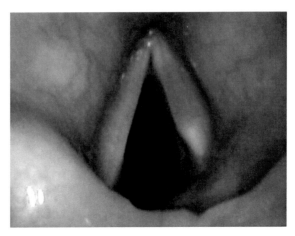

The left vocal cord is longer than the right, so something is structurally different between the two sides of her larynx. This length difference means the vocal cords will tend to vibrate at two different pitches when air passes between them.

I do not know a way of correcting this type of length difference. However, I did put an implant into the right vocal cord pushing it closer to the midline so that she could more completely close the vocal cords. This reduced the air leak and softness in her voice.

Asymmetric summary

Asymmetries between the vocal cords, whether in tension, in mass, in length or in stiffness, will tend to cause the vocal cords to go out of sync with each other during vibration. When out of sync, we will hear two pitches simultaneously. With this double sound or diplophonia, we perceive a rough quality to the voice.

In summary:

- hoarseness is air leaking between two vocal cords that are not closed (white noise) or

- hoarseness is two different sound sources – cords that are asymmetric in some sense (mass, length, tension or stiffness) and vibrating at different pitches, hence diplophonia.

- *Husky hoarseness* is a term that could be substituted for air leak.

- *Rough hoarseness* is a term that could be substituted for diplophonia.

- Looking at the vocal cords during endoscopy, a persistent gap signifies air leak and explains huskiness,

- on endoscopy, the vocal cords should vibrate as mirror images and any deviation from this symmetry can explain roughness,

- disorders may have components of both types of hoarseness,

- hoarseness is a general term that includes huskiness and roughness.

Talkativeness

Vocal swellings

I ask Mrs. Phila Chatterly to tell me about her problem.

"Doc, I talk all the time. I'm off the top of your scale of talkativeness. I can out talk almost anyone. So, it's really important that I get my voice back! You know, I never used to lose my voice and now, every other week, it is almost unusable. I sell real estate. I can't be a good saleswoman if I don't have a good voice ..."

I cut Mrs. Chatterly off mid-sentence with a question, "When did you first notice the problem?"

"Only about six months ago," she replies.

Before she can go on, I ask her if anyone else has looked at her vocal cords before.

"Oh yes, several doctors. The first one said I was normal. The second one said I had reflux and I should stop eating chocolate and coffee, no spicy food, and I was to stop snacking before bedtime. I didn't get any better and when I went back, he said my vocal cords were red and that I should put the head of my bed up four inches and take the pill for reflux twice a day...".

I cut her off again and moved the interview along.

People with a talkative personality frequently associate with other talkative people and talkative people step on each other's conversations without offense. They are used to simultaneous conversation and expect others to jump right in. Unconsciously, they expect that you will just chime in when you have a thought.

I received an education about *listening* to a talkative person early in my laryngology career. I listened intently to Julie Smith tell her very long story about her voice problem. Attempting politeness, I allowed Julie to go on and on about her problem. Julie might never finish her story as far as I could tell. It was over a half hour until I could start my examination. Ultimately, I even had to apologize to my next patient for being so late. At the end of the afternoon, my office nurse informed me that Julie was not coming back. When Julie left, she complained to the nurse that I had paid no attention to her. She said I wasn't listening to her.

That is when I learned that interrupting talkative people will not be perceived as impoliteness, it will be perceived as attention. A talkative person will interpret interruptions to be a sign of great interest. Passive listening to a long story will be perceived as a lack of interest and a coldness on the part of the listener. So I listen carefully to my quiet patients. I listen, but interrupt my talkative patients. I am interested in their story and I also want to direct the conversation toward the details that I need to make a diagnosis. We both get what we want out of the interaction.

It is not just in the doctor's office that talkative people exercise their voice. Even after a talkative person injures their vocal cords, or the cords start to swell from overuse, the talkative person just continues on as if their vocal cords were immortal.

While the vocal cords can vibrate normally, even a million or more times a day, at some point they will start to exhibit wear and tear. This almost always shows up as a swelling in the center of the vocal cord on the edge where they strike against the other side on every vibration. And like the skin on a laborer's hand, the lining of the vocal cord may react to overuse by thickening.

Chronic overuse in terms of quantity of vocal use or in terms of volume or some mixture of volume and quantity may result in a callus forming on the edge of the vocal cord, a vocal nodule in the casual parlance. Most often there is swelling on both vocal cords from this type of overuse. In my experience, talkativeness is usually a lifelong personality trait and the vocal impairment has usually been present for a long time in terms of a husky quality to the voice. At some point the thickening is large enough to impair the voice to such a degree that even the talkative person begins to notice there are vocal limits.

The Bastian scale of talkativeness

When studying with Robert Bastian in 1998, I was introduced to a seven-point scale of talkativeness he had developed. It has proven extremely elucidative in my voice practice.

Both Robert and I enjoy France and travel. When we met, Robert told me a story of travel on a train from France to Italy. He ended up seated in one of those passenger compartments which places six people, three on each side, in close proximity to one other. Beside Robert was an elderly Italian woman talking to her friends. She clearly enjoyed talking and was an accomplished talker, dominating much of the conversation. As the hours passed, this lady's voice began fading in quality. Another person might have rested their voice, but this woman kept talking. In fact, she talked until nothing more than a whisper came out at all when she tried to speak. Robert noted how intrinsically strong the motivation is for some people to talk. They will talk beyond the point where they can even produce sound with their vocal cords.

From this observation that talkative people not only talk continuously, the upper extreme tending to believe they are vocally immortal and don't stop talking even as their voice is fading away, Robert devised a seven-point scale for assessing talkativeness. In this self assessment scale,

- 1 is an innately quiet person,

- 4 is a person of average talkativeness and a

- 7 is a person who lives to talk.

quiet, untalkative			moderately talkative			very talkative
1	2	3	**4**	5	6	**7**

Interestingly when presented with a row of numbers 1 through 7; designed so that 4 is the average person, most people can be rather accurate with their self assessment of talkativeness. A person who can talk until they no longer have any voice at all often rates herself, with a laugh, as an 8 or 10 on the scale. She recognizes the extent of her inner need to talk when asked about it. Much less commonly a partner will chime in and correct her if she tries to underrate her talkativeness.

Most people visiting my office self-rate between a 3 and a 7. To imagine what kind of a person might be a 1 on a 7 point scale, think of a group of farmers gathered in the evening in a room of almost total silence. Ultimately one says something and then after a long pause, another farmer nods assent with perhaps a single syllable affirmation. Quiet people enjoy long periods of silence between speaking, considering the pause to be the normal state of vocal affairs. This tolerance for silence probably proves to be beneficial if one is going to work alone in a field for days at a time and enjoy that occupation.

As a corollary, a person who is a 1 and becomes hoarse, seldom finds himself seeking treatment, since he is seldom using his voice very much anyway. He only comes to my office if someone drags him there.

The scale is meant to address personality, not job, though there is a very high correlation between job selection and talkativeness. It is hard to be happy as a midnight watchman if you are a 7. You would need to be on antidepressants just to survive. Most people who are a 7 find great intrinsic rewards participating in occupations such as acting, singing, performing, teaching, preaching or perhaps just finding a stage to be on. We often, but not always, seek out a job that matches our psychological desire for our degree of innate talkativeness.

This scale is accurate enough that when I reviewed patient charts during my fellowship, I found no patients who had rated themselves a 4 or lower who had vocal cord swellings from overuse.

I use the same 7-point scale for innate vocal loudness.

- 1 is a soft-spoken person,

- 4 is a person with moderate volume and a

- 7 is a person whose voice always projects well.

Volume also has some predictive value for vocal swellings, though not quite as strongly as "talkativeness".

Mrs. Chatterly might have been oblivious to her voice fading away, but if her voice didn't return soon, I bet she sought treatment. For a 7, a working voice is too valuable to do without for long.

Voice & Hoarseness

Summary

Now you know voice is created by vibration and voice can only be produced by vibrating structures. In the larynx the source of vibration is almost always the true vocal cords.

You also know hoarseness is produced in two different ways; by air leak or by asymmetric vibrations. Air leak causes a husky hoarseness. Asymmetric vibrations cause a rough hoarseness. Numerous disorders have both air leak and asymmetric vibrations and perhaps even more than one type of asymmetry.

Air may leak from the front, middle or back of the vocal cords. The vocal cords may be asymmetric in several different ways; mass, tension, length, stiffness or timing.

Two personality traits strongly influence vocal cord disorders. Talkativeness is an innate personality trait that puts a person at risk for mucosal lesions. With more vibrations comes an increasing risk of calluses and swellings which typically leads to a mixed rough and husky hoarseness. A lack of talkativeness puts a person at risk for muscle atrophy and typically husky hoarseness and discomfort from muscle compensation.

Early in this book we considered that the most important part of the question, "Why am I hoarse?" is the *why*. Shouldn't that be obvious? It should be, but it isn't. There is an implicit assumption often made that since voice comes from the larynx, all the examiner has to do is look in the vicinity of the vocal cords and whatever appears on the vocal cords is likely causing the hoarseness. But, mere presence is not sufficient justification. The examiner needs to take this one important step further. The examiner must identify specifically what about the vocal cords is causing air leak or asymmetry. Looking for invisible leaky air is not necessarily intuitive.

The next step in an examination of the vocal cords is to visually look at the vocal cords, typically with an endoscope. I will hold off on the technical details of how to take videos and photos of the vocal cords until later in the book (see "Part III – Examination" on page 253), though if you are interested you could read that section before Part II.

Let's delve into a few examples to see how to use our knowledge to discover the cause and ultimately to decide how to treat various voice disorders. This section will explore hoarseness with more examples from the voice disorders I see in my practice. If you are told you have one of these disorders, you can compare my thought process to what you were told, or what you see of your own vocal cords on a recording. Since the correct diagnosis is the foundation of treatment of the voice, if you are hoarse you need to first assess whether you have been given a correct diagnosis before you start wondering about what is the best treatment or why the treatment you are on does not seem to be working.

Part II – Types of voice disorders

Introduction

Is it vocal behavior or some outside influence that changes the vibrations?

I subdivide causes of hoarseness into two primary categories. I think of these broad categories in terms of what causes the vocal cord dysfunction and so I call them *behavioral hoarseness* and *structural hoarseness*. This categorization is based upon the perceived cause of the hoarseness and ultimately it helps in directing the type of treatment for the problem.

By *behavioral hoarseness* I mean the problem has arisen from *how* the voice is used. The *vocal behavior* or pattern of use causes the problem. An example would be the person who talks so much that the voice becomes hoarse directly from the talking. The overuse causes a change in the structure, usually a swelling arises in the middle of the vocal cord. The swelling allows air to leak around it causing a husky hoarseness. The swelling also creates a stiff portion and increases the mass. These may also create asymmetries and thus a rough hoarseness.

A *structural hoarseness* problem is one in which the vocal cord structure changes as the primary etiology. For example a smoker, over many years of exposing his vocal cords to tobacco ingredients and heat, may develop a cancer. While smoking is a behavior, it is not a vocal behavior. This cancerous growth or bump on the vocal cords has nothing to do with how much or little the person makes the vocal cords vibrate. As the bump in this case actually grows, it changes

85

the weight and stiffness on one vocal cord and causes perhaps both an air leak around the bump (huskiness) as well as two separate pitches (roughness) because of the mass difference.

This distinction between *behavioral hoarseness* and *structural hoarseness* is arbitrary and perhaps even incorrect or imprecise at times. Changes in use sometimes alter the structure and changes in structure may alter the vocal behavior, so they are highly interrelated and deciding which came first might be difficult in some cases.

Nonetheless, categorization of voice problems, based in principle on whether the vocal behavior plays a primary role in the onset of the problem, will prove helpful not only in diagnosing, but even more so in treating the problem.

Behavioral hoarseness

Voice disorders from vocal use can be broken down into three broad subcategories.

- **Mucosal** – movement changes the surface covering
- **Muscular** – movement changes the deeper tissue
- **Nonorganic** – movement inappropriate

Mucosal disorders develop predominantly in vocal overdoers. Vocal overuse traumatizes the surface, that is, the mucosa. The mucosa reacts, so mucosal disorders develop typically from overuse.

Muscular disorders typically develop in vocal underdoers. Lack of use leads to muscle atrophy. People who don't talk much tend to have thinner vocal cord muscles and hence thinner vocal cords. Thinner, atrophic vocal cords leave a central glottic gap allowing air to leak.

Nonorganic disorders are issues with vocal technique. The vocal cords are being closed ineffectively and this pattern of closure may become habitual. Nonorganic voice problems are often associated with stress or with excess tension in the vocal cord muscles.

To define the classification in terms of behavior:

- **Mucosal** – Overuse of the voice

- **Muscular** – Underuse of the voice

- **Nonorganic** – Improper use of the voice

Structural hoarseness

Structural disorders can be broken down into a number of categories. Something extrinsic or intrinsic to the larynx causes the problem, but the problem is not related to vocal behavior. In other words, these disorders are not related to, nor a consequence of how the vocal cords are used.

I utilize the following list as a starting point.

- Congenital

- Inflammatory

- Mucous

- Neurologic

- Trauma

- Tumor

- Unusual (endocrine, hematologic, autoimmune, foreign body)

Let's begin by examining the behavioral disorders and explore the subcategories.

Behavioral hoarseness

Mucosal – Central swellings

Vocal overdoers

Sound production begins when the vocal cords move to the middle and come into alignment parallel to each other. Air is passed through the vocal cords from the windpipe and they start to vibrate, creating sound. As a rough approximation, at a comfortable speaking pitch they open and close perhaps 100 times per second in a typical male voice in North America and perhaps 200 times per second for a typical female. This generates a pitch around C3 on the piano for the male and about G3 for the female.

With each vibration or puff of air the vocal cords collide with each other, most forcefully in the middle of the vocal cord. So if a female talks continuously for an hour, that would amount to 200 collisions per second times 60 seconds per minute times 60 minutes per hour or 720,000 vibrations. It would not take much to have over a million vibrations per day and yet, the vocal cords can generally handle a lot of vibrations. At some point though, the cords collide often enough and strongly enough that the mucosa begins to react.

My daughters love gymnastics. Some nights after working out on the uneven bars they come home and pull the tape off their hands just to show me their calluses and their ripped open skin. They don't get a blister every practice and some gymnasts might never get one, but working out on the uneven bars puts you at great risk for an injury to the covering of the skin on your hands.

Just as the skin of the hands reacts to extensive use by forming blisters and calluses, the same is true for the vocal cords. The skin of the vocal cords, the mucosa, can thicken like a callus, or fluid may accumulate beneath the mucosa in its central portion, like a blister. These are the lesions referred to as vocal nodules and vocal polyps and also called singer's nodules in lay terms. They are no more harmful to the body than a callus on the hand; that is, they are not growths. However, they do have a significant impact on the voice.

Why doesn't everyone get them? There are people who talk non-stop with seemingly few side effects. Other people will lose their soft vocal range after one evening in a bar. Whether due to technique, genetic predisposition or some other factor, some people are clearly more susceptible than others to vocal overuse.

Yet you cannot develop symmetric or nearly symmetric swellings in the middle of the vocal cord other than through vocal overuse. What is important for a diagnostician is the consistency in where swellings from overuse form and the type of vocal impairment they create. They form in the center of the membranous vocal cord on the vibratory margin. They stop the vocal cords from vibrating as the pitch is raised.

There are some less common ways to hold the vocal cords together so that the highest impact area is slightly off-center and swellings may then form off-center. Also when a one-sided swelling occurs, a person may compensate and hold the vocal cords further apart and a reactive swelling, typically smaller, may form on the opposite vocal cord at the striking zone of this initially one-sided swelling.

Swellings from overuse are always benign – not cancerous. They will usually go away with sufficient voice rest, since overuse created them. Three things probably play a role in their development:

1. Quantity of vocal use (talkativeness)
2. Intensity of vocal use (loudness)
3. Technique of vocal use

The quantity of vocal use has the highest correlation with their formation, while volume plays the second most important role. Technique seems to play the smallest role.

Effects of swellings

Since swellings occur in the middle of the vocal cord, they cause several predictable effects. They have mass, so the voice feels heavier. In fact, the pitch does lower as the mass of the cord increases. Because the swelling develops slowly, perhaps over months or years, the patient may not notice this gradual deepening of the pitch. In cases where a thickening of the vocal cord is removed by surgery, the patient will often notice the abrupt lightness and suppleness of the voice after surgery because the change is so immediate. Because the thickening also creates stiffness, the vocal cords do not start vibrating as easily. That is, it takes more pressure from the lungs to initiate vibrations.

When the vocal cords are brought into a closed position, the first thing to happen is that the swellings (they are typically opposing) will touch one another before the vocal cords completely line up, creating a gap in front of and behind the swellings through which air can leak out. Because they touch one another, there will also be a dampening of the vibrations preventing soft phonation.

Think of trying to play the violin while someone lightly touches the string. It will require more force with the bow to set the string vibrating. Because of this touching of the swellings as well as the air leak, the lungs have to push more air through the vocal cords to start the vibrations and there will be audible onset delays at the start of soft phonation.

With time, the individual will develop some compensation. They will start to hold the vocal cords slightly apart to keep the bumps from touching. This requires simultaneous activation of both the

LCA muscle[4] which brings the vocal cords together, as well as activation of the PCA muscle to keep them slightly apart and not touch. Two opposing vocal muscles are essentially competing against one another and the person will soon experience vocal discomfort, pain and fatigue.

Another principle comes into play when we start to vary the pitch. As we increase pitch we increase tension on the vocal cord with the TA muscle and when we shift registers from our chest voice into falsetto, we start to elongate the vocal cords with the CT muscle. Both muscles tighten the vocal cords. As we tighten the vocal cords, any bump in the middle will begin to protrude more and more. Also as we tighten the vocal cords, they become more tense and stiff. If we have a stiff spot in the middle of the vocal cords, the stiffness from the muscle tightening and the stiffness from the swelling add together and at some point for a given lung pressure, the vocal cords are too stiff to vibrate. Even stiff vocal cords though, will eventually vibrate if you place enough air pressure beneath them.

Diagnosis of swellings

I use this knowledge of the effects of vocal cord swellings in order to diagnose them, even at the beginning of an exam. While I am listening to the patient tell me her history, I may hear indications of the presence of vocal cord swellings; onset delays, huskiness, pitch breaks or lower than typical speaking pitch. After the history, I more quantitatively assess a person's vocal capabilities step-by-step (See "Vocal capabilities" on page 265 for a detailed explanation of these tests).

First, I measure the average speaking pitch. Second, I measure the maximum phonation time (MPT) at the speaking pitch. These parameters are only subtly impacted by vocal cord swellings. Third, I establish the lowest and highest pitches that the patient can perform. Since I seldom know what the vocal range was in a given individual

---

4   The full names and actions of the individual muscles of the larynx are described in more detail in the chapter "Intrinsic muscles of the larynx" on page 35.

before the patient had the swellings, I don't get too much information from this range except to document it for the future. Fourth, I ask them to yell. Loudness is not usually impaired. However, up to this point in the exam I have let the patient perform the various tests as loudly or softly as they chose. Now I will limit them to production of soft sound only. At soft volumes, any swelling or gap in the vocal cords will audibly become significantly more apparent.

There are two specific ways to easily hear the phenomenon of vocal swellings impairing soft voice production. I ask the patient to perform a vocal glide (at a soft volume) from low to high pitch. While some individuals may perform this task easily, others, even singers, may not feel comfortable moving slowly up the scale. Another way is to listen to a person sing the first line of the familiar song "Happy Birthday to You." Between the word "day" and "to" there is a jump of three half-steps or three semi-tones. I play a note on the piano for the patient to start on and she (or he) sings the phrase softly.

In the low portion of the vocal range, the vocal cords are short and non-tense, so a swelling does not protrude very much from the cord. Move up a few notes and repeat the test. Raising the pitch, the vocal cords stretch, straighten and tighten. The tightening vocal cord TA muscle pushes the swelling to protrude slightly more. The patient will come to a pitch in her range when the swellings protrude enough to just touch and if the volume is soft, the touching prevents vibrations from starting on the word "to." There will be an onset delay of the sound or it will not come out at all.

If I were to look at the larynx with an endoscope during this test, at this point I would see the bumps touch, dampen or completely prevent vibration and all the air would leak out from in front of and behind the swellings touching each other. This onset delay or no phonation will be on a very consistent note, since below this note, the swellings are apart at the start of vibration and do not impair the initiation of sound.

93

Continuing the test at higher pitches, I may notice a sudden jump in pitch or a double pitch (diplophonia). The bumps have touched each other tightly enough to divide the length of the vibrating vocal cord into two segments, just like touching a guitar string shortens the vibrating segment. The pitch jump can be quite pronounced since the segment that can still vibrate is typically half the length of the vocal cord. Both the front and back segments may vibrate. If the vocal swellings are not exactly in the center, then the front and back segments may be of different lengths and two tones are produced – diplophonia. Depending on how far apart the tones are, it can be quite an unpleasant screech.

If the swellings are quite large, the person may have to hold the vocal cords so far apart that the air leak out the posterior gap prevents the person from having any upper range. One of the principle findings of this test is that the larger the swellings, the lower the pitch at the initial point of vocal impairment.

A patient who is a singer and performing this vocal swelling test every day, will soon learn to identify what vocal activities cause more swelling. For example, she will soon learn that if she went to a loud party the night before, overusing her voice, then the pitch at which soft phonation is impaired will be lower the following morning, signifying there is more swelling from the loud voice use at the party and hence more vocal impairment. If she then rests her voice, she will note over the course of a few days that she can again sing softly at higher and higher pitches, implying that the swelling is becoming smaller in size. This daily performance of the swelling test provides immediate feedback that can then modify future behavior.

If she is a vocal overdoer by history and the vocal capabilities point to an abrupt point in the range where there is a vocal onset delay, then the physician will almost certainly find on endoscopy, a swelling in the middle of the membranous vocal cords on one or both sides.

Earlier I said the swellings are nearly always symmetric. If there is a rule, there are likely to be exceptions. Sometimes there is a swelling on only one side. It is usually slightly translucent (a polyp) and the history may point to a single date where the patient said, "I was fine" and then he yelled and lost his voice. The patient may not even be a chronic vocal overdoer. This injury is more of a vocal indiscretion. Though I have never seen someone the day of such an event, I surmise from the history he may have popped a blood vessel on the vocal cord from his vocal effort. This creates a bloody swelling on the edge of the cord that the body never reabsorbs. When I have surgically removed these one-sided, translucent lesions (something like a blister), they are filled with a jelly-like material, presumably leftover proteins from the blood. The audible effect of this kind of swelling on the vocal capabilities testing is the same as for bilateral swellings. Increasing pitch, on some note, the one-sided swelling touches the other vocal cord and stops soft vibrations.

With an endoscope, I typically see dilated blood vessels, tortuous blood vessels or blood vessels that run more perpendicular across the vocal cords rather than parallel to the edge of the cord. Since vocal cord mucosal swellings are caused by vocal trauma, it is not surprising this same individual may have ruptured the small vessels in the vocal cords many times previously. When these ruptured vessels heal, the person may be left with a tiny aneurysm or the vessel walls may be weak, dilated and tortuous.

When new vessels grow into a place of injury they seem to come from the shortest direction, that is from the side rather than from the end of the vocal cord. If these vessels are right on the edge of the vocal cord, they will be thrown around with every vibration, filling up with blood and dilating with voice use. With voice rest the blood empties back out and the swelling goes back down in size. This can result in the complaint that a singer can perform well for a short time, but then the voice rapidly becomes impaired, only to improve again after even a short rest. The filling up with blood makes the swelling a dynamic swelling.

In summary, a history of vocal overuse, either acutely or chronically, followed by a vocal capabilities exam that demonstrates onset delays and upper range impairment with soft singing should orient the examiner to find a vocal swelling in the middle of the vibrating vocal cord. We will mention some treatments in a later sections.

Nodules

Paired, hard bumps

During Mrs. Chatterly's interview and history, I note a husky quality to her speaking voice.

I see how long she can make a sound and then although she is not a singer, I eventually persuade her to say "eee" on various notes and we work our way down and then up in pitch. At a soft volume, there is a very consistent note on which her voice cuts out. If she sings loudly, she can generate that note, but not when she sings softly. As she tries to sing the note, air comes out and then the note starts – there is an onset delay. Her pitch will also suddenly jump very high; a pitch break. She perceives the sound as a squeak and, embarrassed, stops. At some of the higher pitch attempts, two different, high pitched notes are produced simultaneously. This diplophonia is quite clear as the vibrating sections are completely separate from each other.

I expect to find vocal swellings on endoscopy and these high notes are the ones at which we will see the vocal swellings touching each other on a stroboscopy exam. When I look, I find the expected bump in the center of each vocal cord.

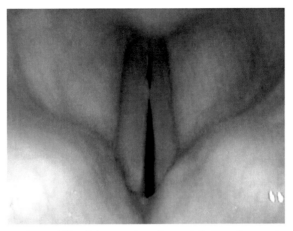

Nodules in the middle of the vocal cords touch before the vocal cords can completely close. This touching is most noticeable at the high pitch here because the vocal cords are stretched tight causing protrusion of the bumps.

Nodules seem to typically form over a long period of time as talkativeness most often is a lifelong personality trait. On occasion, they may form in a relatively brief period of time.

Paired, soft bumps

Johnny Quicktemper and his wife weren't getting along all that well and when she announced that she was moving out, he took to yelling at her. Seemingly, all their conversations ended in yelling or screaming.

He developed hoarseness and a sore throat. He believed the hoarseness represented an infectious laryngitis because his throat was sore and requested an antibiotic from his primary care physician. His voice didn't improve on the antibiotic. (Johnny failed to mention to his physician all the screaming going on in his home; embarrassing perhaps, but likely he also didn't suspect that the increase in screaming might be causing his hoarseness).

When I examine him, there are two white thickenings, one in the center of each vocal cord, directly on the edge.

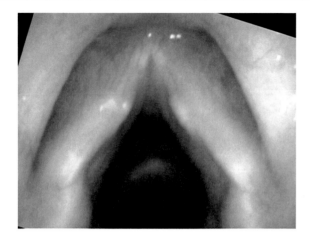

Large nodules viewed on the edges of the vocal cords during breathing. The central part of each nodule is white because as mucosa thickens or forms a callus, you can no longer see through the mucosa to the underlying structures.

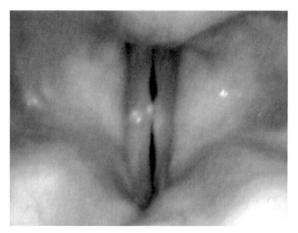

When the cords close to make sound, the white thickenings touch in the middle, leaving a gap on either end to leak air or if enough pressure is used the two short segments vibrate at high pitches.

Johnny's pain is not from infection. It is from all the tension he is using to keep the vocal folds far enough apart to allow them to vibrate so he could continue his job in sales.

Tension keeping the vocal cords apart allows air to leak. The calluses also make the vocal cords stiffer, requiring more airflow to keep them vibrating. This increase in airflow along with the larger gap lead to more turbulent airflow through the vocal cords and gave his voice a prominent husky quality.

The laryngitis quality to his voice is augmented because the vocal fold calluses make the vocal cords heavier (which would by itself only lower the pitch), especially because they became asymmetrically heavier. They end up vibrating at two different pitches giving his voice a roughness (diplophonia). This compound hoarseness of huskiness and diplophonia are very similar to the sound quality of infectious laryngitis. The recent onset of the hoarseness along with the pain in his throat is compatible time wise with acute infectious laryngitis. The discreet note where Johnny experiences an audible onset delay and the endoscopic exam are not consistent with acute infectious laryngitis.

In just over a month of voice rest (and some marital counseling – they can still irritate each other, but without the screaming), his voice improves to the point where he can again talk all day in his sales position. His pain resolves. Even though the swellings are still present to some degree, they no longer significantly impair his speaking voice and as a salesman, that satisfies his vocal needs.

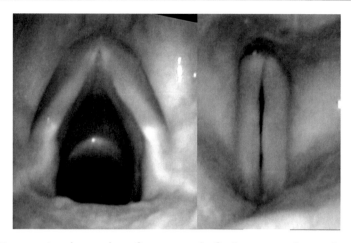

Comparative photos taken after one month of voice rest or at least reduced vocal use. The swellings are still present, but reduced in size. The vocal cords can come closer together with less air leak (right).

Swellings don't have to go completely away to be a success, they just have to go down enough in size that the patient can maintain their needed level of vocal function. For the Mozart soprano, swellings may still impair the voice at almost microscopic sizes. For a salesman, they may tolerate – indeed not even notice – rather moderate sized swellings. At the other end of the spectrum, a heavy metal singer might not sound as good without his swellings.

Swellings that increase in size over a few days or weeks will likely resolve in days or weeks with a decrease or moderation in vocal use. Swellings that come on over years will likely take much longer to reduce in size with vocal rest or vocal moderation.

Voice rest

What is voice rest?

Not making a sound. Easy enough.

Not really.

First of all, people who acquire vocal swellings almost always love to talk. Their inner batteries are recharged by talking with others. Turn off the conversation and the person is bummed out. Absolute voice rest is incredibly difficult in a psychological sense.

Second, most talkative people find ways to fill their lives with talking. They have an active social calendar. They use their mobile continuously. They find a job completely dependent on talking. Silence in this type of job doesn't pay. Talkative people don't become computer programmers that sit up all night in front of a monitor. They gravitate to acting, sports coach, vocal performance – any activity that allows for a lot of vocal use.

Thirdly, a talkative person judges the vocal behavior of others relative to themselves. They believe that the amount of talking they do is "normal" and any amount less is abnormal. Ms. Chatterly, a 12-14 hour a day talker, might well view 6 hours of talking a day as "absolute voice rest." It is almost impossible for the talkative person to rest their voice for weeks, much less months.

So, in practice, short trials of voice rest may be helpful for recent onsets of hoarseness, but more likely, there needs to be a long-term approach to modification of vocal use. Results will be somewhat slow to come. There will be relapses of vocal overuse in the chronic talker. Vocal modification takes a significant effort and commitment.

Swellings may be peaked or broad

Sometimes larger lesions are more difficult to visualize

Mrs. Chatterly's vocal swellings were fairly discreet. Peaked swellings are the easiest to see. They stand out from the edge of the vocal cord like the volcano, Mt. Hood, viewed in the distance from my home in Portland. They are obvious. Johnny's swellings were a bit broader, but still discrete enough to be easily visible on endoscopic exam.

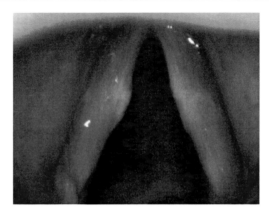

Discreet peaked swellings resembling mountains. The apparent "snow" on the swellings is callus – thickened skin.

The patient quickly learns to hold their vocal cords farther and farther apart to accommodate enlarging swellings. This may increase the degree to which the central portion of the vocal cord continues to be traumatized during sound production. The farther the cords are held apart, the more traumatic the impact when the cords strike each other, which may lead to these peaked type of nodules. Long-term vocal overuse may alternately produce thickening along a rather broad portion of the length of the vocal cord.

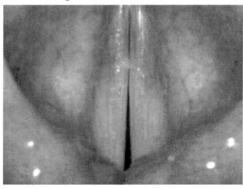

Broad-based swellings, viewed from above (here with the rigid endoscope), may not look like swellings at all. All that seems apparent is the vocal cords are open at the back. There is a little mucous on the center of the vocal cords.

Broader-based swellings, even though larger in mass, are more difficult for physicians to perceive. The very talkative person has underlying TA muscle hypertrophy – basically they are a vocal body builder. This overall muscle thickness is spread along the entire length of the vocal cord. Then on the surface, the mucosa has likely thickened from use as well.

Secondly, at typical speaking pitches the TA muscles are quite relaxed and so the surface thickening subsides into the substance of the vocal cords during vibration. The thickening is again more readily apparent as the person raises their pitch, lengthening and tensing their vocal cords. At high pitches, TA muscle tightness forces the swelling to stand out from the straight line that would otherwise represent the edge of the vocal cord. As the swelling stands out from the vibratory axis of the cord, the vocal processes at the posterior edge of the vibrating cord will need to move farther apart to keep the edges of the thickened cords from touching even with broad-based swellings. Otherwise, touching will still stop the vocal cord vibrations. The muscle tension holding the vocal cords apart is not primary, it is a necessary compensation and secondary to the swellings (and can cause significant, generalized neck pain).

Thirdly, these vocal cords are broad not only along the length of the cord, which is what is easily viewed from above (the physician's typical rigid endoscope perspective), but the cords are thicker in diameter. You can hear this thickness as a deeper-than-usual pitch, but if you are meeting a person for the first time, how do you know what her usual pitch used to be? If a flexible endoscope is placed very close and flexed so that it is more parallel to the length of the vocal cord, these thickenings can be better appreciated.

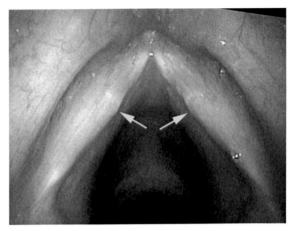

When the flexible endoscope is inserted in the same patient as the previous photo, and angled to view the cords from posterior, the central swellings are more apparent (arrows point to the center of thickenings) Additionally, the thickenings extend in all directions: anterior, posterior, superior, inferior.

What appears to be only a rather small posterior gap when viewed from directly above with a rigid endoscope actually represents a significant change to the vibrating edge of the vocal cord in terms of increased mass and increased stiffness and the mass is better appreciated from a posterior view. Larger swellings, along most of the length of the vocal cord, can be more easily missed, while smaller swellings are easier to see. If you can hear a vocal impairment, the pathology may be right before your eyes, yet so easy to miss!

Swellings may be tiny and hidden

Keep looking until you find what you can hear.

Amour T. Cher gives me a classic history for vocal nodules. She is an non-stop talker – 7 on the Bastian scale of talkativeness. Amour found a job as a high school teacher, so she has an opportunity to talk every day. For the past six years she has lost her voice completely about eight times per year, almost once a month during the school year. This aphonia (*a-* – without, *-phonia* – voice) lasts for three to five days and when her voice comes back again, it comes out in squeaks and puffs of air. The more she uses her voice, the worse these episodes. This history suggests vocal nodules or at least a swelling on the vocal cords until proven otherwise.

During Amour's exam she has a hard time controlling her vocal pitch. Not everyone is a born singer, so it is possible that she merely has poor vocal control. However, a few times when she attempts a high pitched voice, I hear a squeak – an abrupt jump upward even higher in pitch. Remember, when vocal swellings touch, the vibrations suddenly go from the whole vocal cord to half or less of the vocal cord so the pitch jumps correspondingly, typically going quite high.

I expect to see vocal swellings based on her talkativeness and the squeak I am hearing, but when I put the endoscope in, her vocal cords look entirely normal. I am surprised.

I have pretty good equipment and still it happens that at times everything at first appears normal. In my reasoning, there are two possibilities: I am either not seeing the pathology despite its presence or I am misinterpreting something from my history or vocal capabilities portions of the exam.

Certainly with the typical endoscope in the typical physician's office in this day and age a view from high in the back of the throat is going to miss a great many problems on the vocal cords. The cords are small and far away on the screen. There is pixilation of the image if fiber-optic technology is used. The vocal cords move quickly. Many patients are gaggy. There are many reasons to miss small lesions.

Yet, I have a very high-tech, digital chip endoscope and I am able to put the scope reasonably close to the vocal cords without making most people gag. I still couldn't see any lesions on Amour's vocal cords.

I think back to her history. On the one hand, during this recent summer vacation she hadn't lost her voice so maybe she had temporarily improved. However, I had heard the problem in her upper vocal range while listening to her, so I felt the more likely explanation was that I was missing something on my endoscopic exam.

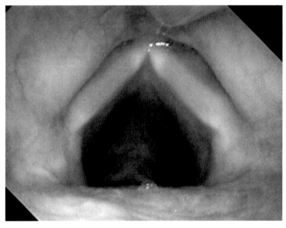

Reasonably close view with a high-tech endoscope and apparently normal looking vocal cords, at least from this mid-range distance and perspective.

I sprayed Amour's throat and her vocal cords with lidocaine to anesthetize them. I then passed the endoscope close enough to touch her vocal cords. The lidocaine prevented any gagging. I had her sing a note and then work her way up the scale until she reached the higher notes. When I heard the squeak again – the pitch break – I could see the problem using the endoscope.

Even though the vocal cords grossly act like a string, they are thicker than most strings and there are actually two lips on the edge of the vocal cords. When the vocal cords oscillate, the upper lip mucosa of the cord touches first and then separates revealing the lower lip. The last bit of mucosa pulling apart is the lower lip.

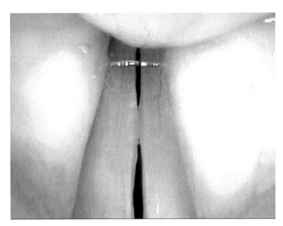

At this high pitch, very close range, during stroboscopy, as the upper lips of the vocal cords have just pulled apart, the central swellings on the lower lips of her vocal cords are visible touching each other and seem all too obvious. The two air gaps created are very apparent in this view as the dark slits.

Her swellings (nodules) were small enough and they were located on the lower lip such that they were covered up most of the time. Only when my endoscope was almost sitting directly on the vocal cords and when the patient was at her highest pitches could I see the swellings and watch them break up the vibrating wave.

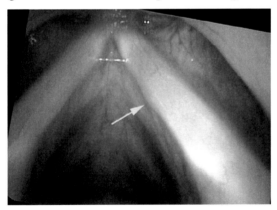

Maneuvering the endoscope to an ultra-close view, here almost parallel to the vocal cord, we can see under the edge of the vocal cord. The swelling on the bottom edge or lower lip of the right vocal cord (arrow) is clearly visible.

107

Ultimately some combination of a high-quality video recording, combined with endoscopic technique, combined with a stroboscope, combined with vocal manipulation will produce answers for seemingly invisible lesions. Closeness matters for an endoscopic exam. Pitch matters for an endoscopic exam. Volume matters for an endoscopic exam. If you can hear the problem, you can (and must) see it as the diagnosing caregiver. The patient's history, the audible exam and the visual exam should all correlate.

Polyps

Vocal cord blisters, acute overuse and hemorrhagic polyps

Charles Earl Osterman manages a large medical group. He is not naturally chatty (a 4 out of 7 on the Bastian talkativeness scale). However, he needs to be trusted when he speaks and a recent onset of hoarseness is interfering with his work. An otolaryngologist told him he had a nodule and after a week of voice rest (presumably no talking at all) he was sent to speech therapy for a month. He really doesn't notice any improvement.

I hear the roughness and the air leak in his voice when he speaks and these qualities are even more evident when I test his vocal capabilities, especially on the swelling tests. On endoscopy, I note a one-sided, spherical swelling on the vibrating edge of his vocal cord with blood in it.

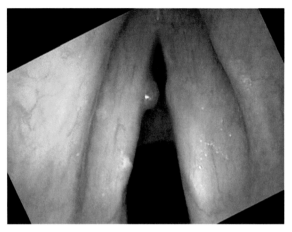

The left-sided swelling is a blood filled polyp, red in color and hemispherical. There is a smaller reactive swelling on the right side. That is, the polyp on the left strikes the right cord hundreds of thousands of times in the same spot and induces another swelling.

There is a large and slightly tortuous blood vessel crossing the surface of the vocal cord and leading to this polyp. The polyp flips up and out of the way when he speaks at a low pitch. If I ask him to raise his pitch, the polyp stands out and prevents him from making a higher pitch. He doesn't sing, so it doesn't bother him too much that he can't reach high notes and explains why he could tolerate such a large swelling before seeking treatment. On the other vocal cord there is a small (and I will suggest, reactive) swelling.

One-sided (or highly asymmetric) lesions are usually found in vocal non-overdoers. I see two typical types of one-sided lesions. The clear, translucent, mucoid fluid-filled polyp and the more spherical blood vessel-fed lesion, partially or completely filled with blood. There may be a swelling on the other side, but it is typically an order of magnitude smaller.

I suspect that one-sided polyps (with a feeding vessel or without) are commonly caused by a single vocal indiscretion such as a scream at a football game or perhaps a single vocal trauma such as a cough. A loud, forceful slamming of the vocal cords together causes a rupture of a blood vessel. The blood beneath the mucosa causes a raised area.

In the case of a clear polyp, the blood has reabsorbed, but a proteinaceous, mucoid fluid remains. In the case of a broken blood vessel, it heals with a thin wall, effectively creating a weak aneurysm on the edge of the vocal cord that is repeatedly exposed to the trauma of phonation. That blood vessel often persists in a dilated form.

It may keep breaking and re-filling the polyp with blood. There may be some tendency for the user to try even harder and louder to get a sound out, further traumatizing the swelling.

The voice may worsen temporarily with vocal use because the polyp flipping around on the edge of the cord centrifugally fills with blood and temporarily enlarges. With voice rest, the blood drains from the polyp and the voice returns to its baseline level of hoarseness.

In Charles' case, the vocal polyp is red, so it is filled with blood. He has continued to speak with this swelling on the edge of his vocal cord and that has led to a reactive swelling on the opposite side.

Since I doubt, in this type of one-sided lesion, that there is a chronic vocal behavior problem, my primary approach is to surgically remove the lesion and try to cauterize or interrupt the feeding blood vessel. I don't start with any vocal behavior modification as I can't identify a bad habit to be modified.

After surgery, Charles Earl Osterman was vocally improved and returned to regular work. Three months later with a cough, he developed a hoarse voice again. The hemorrhagic polyp was back. Although this was the first time I had seen a recurrence, the elevation on the opposite vocal cord was gone. There was a new polyp, again with blood in it and a feeding blood vessel was again present.

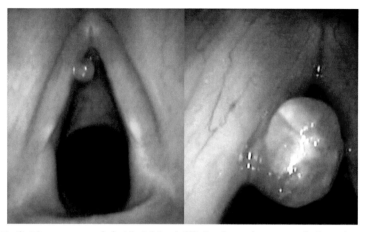

Left: *The recurrent-left sided, blood-filled polyp, when viewed ultra-close, seems to be on a stalk. There appears to be a "feeding" vessel adjacent to it.* *Right*: *The polyp is lumpy from blood clotting in it.*

I again removed the polyp, this time in the office. I cut it off at the base and cauterized the feeding blood vessel with a laser.

It seems to me that if an enlarged vessel is present on the vibrating edge of a vocal cord, it will be more subject to breakage from vocal trauma. I suspect that I must not have completely cauterized this vessel at the time of the first surgery, there was a second vessel I missed or a new vessel regrew after I initially ablated it.

The polyp has not recurred in over a year since then, though it is plausible that he could break another blood vessel in the future. Since it has not recurred in a similar three month time window, it seems likely that the offending vessel is no longer present.

Smoker's polyps

Tobacco & Talkativeness

Puije Parlet Fume has had trouble with a husky voice since childhood and she is now in her mid-sixties. Her huskiness has become worse over the past year and a half and her voice not only varies in quality, but at times it fades completely out with use. Sometimes her voice suddenly chokes off and she can't generate a sound. Speaking is becoming more effortful. She finds it almost impossible to yell. Recently she has begun to notice increasing shortness of breath.

She has been smoking about two packs of cigarettes a day for the past 50 years, though she is now trying to quit and has started a medication to help her. She has been talkative all her life and rates herself as a 7 out of 7 on the talkativeness scale.

Listening to her voice, her pitch would be considered low, even for a man, although she doesn't mind her deep pitch. It is quite rough as well. All her friends recognize her voice for what it is.

On her endoscopic examination she has large polyps extending the length of both vocal cords. They are based on the superior surface and they move in and out with speaking and breathing. At times they get caught below her vocal cords and no sound comes out at all. Because they are so large, at times she squeezes her false vocal cords close enough together and makes them vibrate.

Because of the nearly universal association with smoking tobacco, this type of polyp is known as smoker's polyps. The other common name is Reinke's edema, named after an early article about them.

The vibration of the false vocal cords contributes both to her deep pitch and the rough quality of her voice. Ordinarily the false vocal cords are very thick relative to the true vocal cords. They always create a very low pitch if they are used as a sound source. Additionally, each of her true vocal cords is a different size, the asymmetric mass of the true vocal cords adding in two additional pitches. She is making up to three pitches at one time and that accounts for her very rough voice.

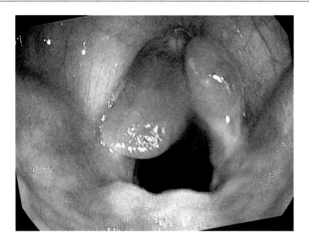

This angled view from behind the vocal cords shows the attachment of the polyps on the top surface of the vocal cords.

The stiffness of the polyps and the false cords produces the effortfulness she senses while trying to speak. With this stiffness she also tends to channel some of the air toward the posterior, non-vibratory portion of the vocal cords. This air leak creates the husky aspect of her voice.

While smokers can have several reasons to be short of breath, when the polyps are as large as Ms. Fume's polyps, they become one significant contributing reason. There may even be noise created when breathing in, if the polyps are close enough together to vibrate on inspiration, medially called inspiratory stridor.

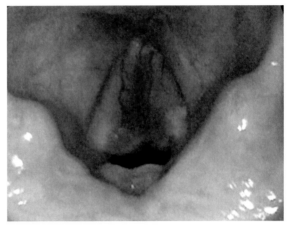

These polyps are based on the top surface of the vocal cords. When she breathes inward, they suck together, touch each other and obstruct much of her airway. This contributes to her sense of shortness of breath when trying to inhale.

Smoker's polyps have quite a few ectatic blood vessels as visible here. It is possible that the heat or chemicals from tobacco smoke activate this blood vessel growth and this extra blood supply may be what contributes to the formation of the polyps. These blood vessels also allow the polyps to be treated with lasers that target blood vessels such as the pulsed KTP laser.

She was planning to quit smoking and desired an easier to use voice so she elected to undergo a surgery to partially excise these polyps. They were removed partially by cutting and partially with KTP laser ablation.

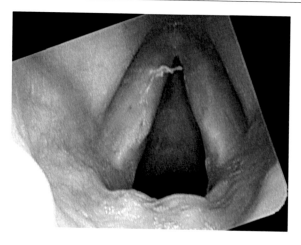

Vocal cords after excision and KTP laser treatment. The vocal cord margins are straight and the mass of the vocal cords is a more normal size. The white spot on the left vocal cord is an absorbable suture still present after surgery.

Ms. Fume quit smoking on the day of surgery. Her pitch rose up about half an octave by two months after the treatment (it takes some time for the swelling of surgery to resolve). Her vocal cords are now a more normal size. Her voice is still rather low for a female, but she no longer needs to use her false vocal cords to make sound.

If a person smokes, but is not chatty (not a 5, 6 or 7 on the talkativeness scale), he or she will not form smoker's polyps. If a person is talkative, but doesn't smoke tobacco, he or she will not form smoker's polyps. Both conditions – talking and smoking – are required to develop smoker's polyps.

Quitting smoking will not generally reduce the size of the polyps though it may stabilize their size so they don't continue to enlarge. Quitting talking seems never to be an option for my talkative patients. Consequently, I typically only surgically reduce their size if the person does desire a higher speaking pitch (and they are usually trying to quit smoking). Since most individuals with smoker's polyps don't mind their pitch, the other main reason I treat smoker's polyps with surgery or laser is if they are impairing a person's breathing.

Voice therapy

Unilateral vs. bilateral swelling, nodule vs. polyp

What is the point of making distinctions between nodules and polyps? Is there a reason to differentiate between unilateral and bilateral swellings?

A nodule represents a thickening of the mucosa, the outer layer of the vocal cord. It is analogous to a callus on the hand.

A polyp is more like a blister. A polyp has a fluid collection beneath an otherwise normal outer layer of the vocal cord.

To the patient, the vocal impairment is the same. To a surgeon, treatment timing should be different.

As mentioned above, I have found that polyps, particularly one-sided polyps, and especially one-sided hemorrhagic polyps, tend to be from vocal accidents or vocal indiscretions. Often the history will corroborate the sudden onset. Quite often the person is not even very talkative, so it doesn't make sense to modify or further decrease vocal use. The person's daily pattern of voice use doesn't need significant management, though one may want to avoid screaming at a football game. I have had patients who chose to try long periods of voice rest anyway and voice rest doesn't tend to dramatically change the size of polyps. It seems reasonable to intervene early with a surgical excision and generally, the polyp does not tend to recur after surgical excision. Consequently, I recommend surgery as the primary approach for fixing one-sided lesions, particularly polyps and especially hemorrhagic polyps (since the blood vessels may rupture again with even moderte vocal use).

At the other end of the spectrum, bilateral vocal nodules in a vocal overdoer seem to reappear quickly after an excision unless there has been significant management of the person's vocal overuse. So voice therapy designed to give the patient insight into how much they are really overusing their voice will diminish recurrences (remember, a vocal overdoer, a 6 or a 7 believes that non-stop, all day talking is normal vocal behavior). If behavioral management succeeds in mitigating overuse yet fails to shrink the lesion, surgery becomes an option.

After surgery for vocal swellings, voice therapy may again be quite helpful to prevent recurrence of vocal nodules. After removal of these swellings, the voice typically becomes very easy to use, and that proves to be a license to talk for the vocal overdoer. A good voice therapist may be able to modify the patient's vocal behavior.

Muscle memory also has a significant impact on the voice. If the vocal swellings have been present a long time, the post-surgery patient may continue to hold the vocal cords apart out of habit and air will still leak, even though no swellings are present. With the vocal cords apart, they may be striking each other more traumatically than if they were positioned against each other. A good voice therapist may be able to induce bringing the vocal cords together into apposition before producing sound and may be able to reduce the impact of each vibration by altering breath support and resonance. It is possible to generate loudness by increasing airflow through the vocal cords (a vocally traumatic method of increasing volume), but also by matching the resonance of the pharynx to the vocal cords (an efficient and vocally sparing method of increasing volume).

Mucosal – Ectasia

Abnormal vessels: dilated capillaries, tortuous capillaries

Tiny capillaries run the length of the vocal cords. They run parallel to the edge of the vocal cord. When a vocal cord is traumatized – by chronic over-talking, high-volume talking, screaming, a cough or aggressive singing, for example – blood vessels may break and blood seeps out into the cord. This results in a beet red vocal cord after an acute injury. As the blood absorbs over several weeks, the vocal cord appears bruised and turns gradually more yellow before all the hemoglobin is reabsorbed by the body.

The broken blood vessel, as it heals, will often take a more tortuous path or remain dilated, perhaps because the injured walls of the vessel are thinner. So whenever I see unusual capillaries – capillaries taking a tortuous path or capillaries running at odd angles to the length of the vocal cord – I suspect a prior history of severe vocal overuse. These unusual vessels are often located in the mid-portion of the vocal cord where the most trauma occurs.

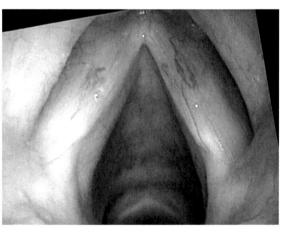

Dilated, tortuous capillaries in the middle portion of the vocal cord. They form a historical record of vocal overuse.

Sometimes capillaries are located on the edge of the vocal cord and enter a polyp. When they are located on the vibratory edge of the vocal cord, they can be quite troublesome. As the vocal cords vibrate rapidly these vessels become filled with blood and swell, impairing the vibrations of the vocal cord and the person's voice. In this location they are more likely to rupture with further vocal use.

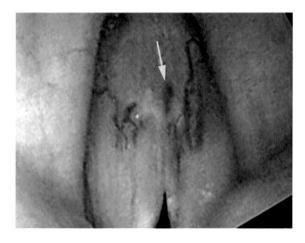

During stroboscopy, the vibrating edge of the vocal cord everts and dilated capillaries on the edge of the right vocal cord become visible as they rotate upward.

Another type of trauma is radiation therapy. After radiation is used for treatment of vocal cord cancer very tortuous capillaries grow gradually over several years in the radiation field. They resemble roots you might see as you lift a plant out of a pot. As long as the capillaries are not on the edge of the vocal cord, they generally don't interfere with the production of sound. When they form on the edge of the vocal cord, they may dilate with vocal use and impair the closure of the vocal cords.

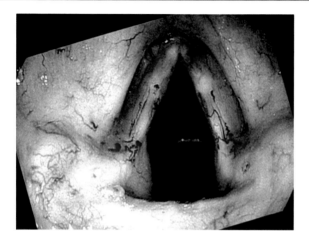

Capillary telangiectasias are visible all over the larynx, both on the true vocal cords and on the false vocal cords, secondary to radiation therapy about 15 years ago. A few of these capillaries are on the vibrating edge of the vocal cord and are causing some hoarseness by impairing vocal cord closure.

Mucosal – Granuloma

Trauma to the vocal cords, inflammatory healing

"My ENT doc sent me here," Sam Sales notes. "I went in to see him because of a pain in my neck on the right side about six months ago. He looked with a scope and found a growth on my voice box. He took it off and it was benign. However, on a check up, it came back, so he put me on an antacid[5]. I didn't have any heartburn, but he said silent reflux can come up and burn my voice box and cause a growth. The pain has gone away, but I am worried about cancer on my vocal cords because I was exposed to a lot of second-hand smoke when growing up."

Sam uses his voice all day as a consultant.

"My voice fades out by the end of the day. It usually recovers by the next day," Sam notes.

I was involved in an accident three years ago and had a tube put down my throat. That may have been the cause of this," he speculates.

"Did your voice change when they took the tube out three years ago?" I ask.

"No, it was the same. My voice only began changing about six months ago."

On his medical report, his otolaryngologist had removed a granuloma. It had come back rather quickly. The powerful proton-pump inhibitor had no effect on the recurrence. On endoscopy, there was a surprisingly large lump on the right vocal process.

---

5   Dexlansoprazole – the newest medication for GERD at the time.

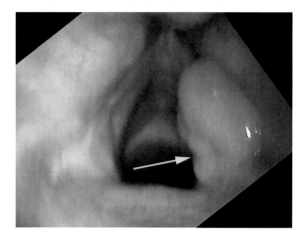

A large yellow ball of tissue is located just above a right vocal process ulcer (arrow points to groove of ulcer).

There is one particular type of vocal cord lesion that may start out with a temporary pain. The lesion will typically be located somewhere on the back of the larynx. The most common location is on the cartilage called the vocal process of the arytenoid, right at the end of the vibrating portion of the vocal cords, but it can occur on the arytenoids wherever cartilage closes against cartilage.

It really isn't a growth, rather it is a collection of blood vessels wound into a little ball. Similar tissue is called "proud flesh" if it is on the outside of the body. When something traumatizes the mucosa, eroding through it and exposing the cartilage, an ulcer occurs. Then granulation tissue or a granuloma forms as a reaction to an open wound exposed to the bacteria in the mouth and throat.

The open ulcer often initially causes pain. If you have ever had an ulcer in your mouth, you know the character of the pain. It can be quite sharp. The pain can masquerade as ear pain because the tenth cranial nerve supplies not only the larynx but also the ear. It is not infrequent for individuals to experience pain shooting up into the ear on the same side. This pain, one type of otalgia, is called a "referred pain" and an ear doctor should always be checking the vocal cords when there is a complaint of ear pain because of this pattern. The pain is often worse after talking since it traumatizes the area and may reopen the ulcer.

Pain from an ulcer is different from muscle tension pain. Muscle tension pain generally has an achy or burning quality and is located in the mid-line or all over the central neck. It is diffuse. Pain from an ulcer is one-sided and can be quite sharp. It generally feels quite localized.

The reason for Sam's lump was a traumatic wound of the larynx that created an ulcer. How did Sam injure himself? How do we treat the condition?

The answer to the first question will determine our response to the second. Trauma can be a single instance or it can be recurring.

Intubation injury

An example of a single injury which might cause a granuloma is from an intubation. During a surgery under general anesthesia or after an accident, an anesthesiologist or emergency medical technician places a plastic tube through the larynx, between the vocal cords into the trachea, through which the anesthesiologist supplies oxygen to the lungs. That endotracheal tube tends to put the most pressure on bumps or prominences at the narrowing of the airway. The vocal process on each side is one such bump that stands prominent. After the endotracheal tube is removed, quite often an ulcer forms on or near the vocal process.

While these ulcers are thought to be rare, I suspect they are more common than believed. Frequently, I examine some of my patients with an endoscope every day after a general anesthetic. I find ulcers occur rather commonly in this group. In all of these cases the ulcer or ulcers form and heal within a few days without any residual visible problem. None of these ulcers developed into granulomas, but this suggests it is from a lack of looking that ulcers from endotracheal tubes are thought to be uncommon. If ulcers are common, then granulomas might be more common than believed since seldom does anyone look.

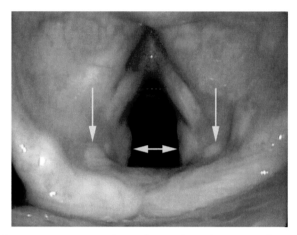

Four ulcers are present two days after a relatively easy intubation by an anesthesiologist (on another patient, this is not Sam). These healed within a week.

Clearly, sometimes a granuloma forms from an injury to the mucosa and becomes symptomatic. The granuloma may form on one or both sides. It typically enlarges for several weeks or even months and then develops a stalk and starts flipping in and out with breathing and speaking. Subsequently, it usually breaks off at the stalk and is coughed out. If an intubation granuloma does not fall off quickly enough, it can be surgically removed.

Sam's injury was likely not from an endotracheal tube. He wasn't hoarse until several years after his intubation. I suspect his granuloma is a result of chronic trauma. In support of this theory, when he had the granuloma removed, it left a small raw area where the stalk was cut. He resumed talking after the surgery and continued to traumatize the vocal process, so the ulcer and granuloma recurred. The underlying cause – chronic vocal process trauma – was not addressed.

We watched the granuloma for awhile and it enlarged and became smooth. His work preventing him from stopping talking completely. I can enforce voice rest by injecting botulinum toxin into the vocal cord muscle and the ulcers and granuloma will go away, suggesting cessation of trauma will cure this condition.

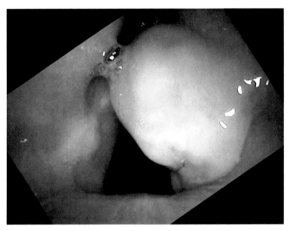

The right granuloma is now white, smooth, mature and on a stalk. I can flip it around with my endoscope. There is also a grooved ulcer present on the left now.

In Sam, I injected a steroid (triamcinolone, a strong anti-inflammatory) into the vocal cord granuloma in the office with the goal of reducing the size of the granuloma. I also injected botulinum toxin into the LCA muscle with the goal of weakening closure of the vocal cords and reducing the trauma on the vocal process cartilage from making sound. When he returned in two months, with his symptoms improved, rather than having coughed the granuloma out, as frequently happens, the granuloma had shrunk.

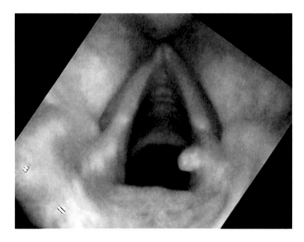

Two months after steroid and botulinum toxin injection, the granuloma is reduced in size.

Chronic trauma of the vocal process occurs more frequently in people with a vocal cord weakness. A weak, thin vocal cord, whether from a nerve injury or from lack of use, thins and bows as the thyro-arytenoid muscle within the vocal cord atrophies. This person then squeezes the vocal cords more firmly together to compensate for the bowing and the vocal process cartilage ends up striking the oppos-ing cartilage more firmly. This is Sam's issue. His vocal cords are thin (photo above). When he closes his vocal cords together, they are very bowed and the vocal process, where his granulomas are forming, strike each other.

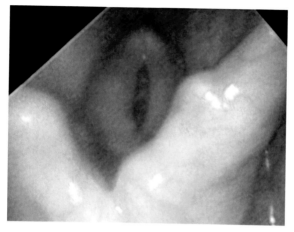

The moment of onset of closure. The vocal cords are very bowed. The vocal processes strike each other with each closure of the vocal cords.

Another chronic trauma I encounter is a prolonged cough. The vocal processes traumatize each other from repeated very forceful closure. A severe vocal overuser, especially a loud vocal overuser, aggressively closes the vocal processes against each other. Underlying all of these conditions is the forceful and possibly frequent striking of the vocal processes of the arytenoid cartilages against each other. The mucosa overlying cartilage erodes and an ulcer is formed.

On endoscopy, new ulcers from overuse, are initially split into two parts where the opposite vocal process rests in a groove that it creates during closure. Gradually, granulation tissue heaps up around the traumatized point, forming a mass or possibly two, that coalesce into a granulomas.

In the middle of their lifespan, granulomas become yellow or red and round. Later, this mass of tissue tends to harden, become white and spherical and the base narrows to a stalk. They evolve from an ulcer to an active granuloma and then to a mature granuloma finally, which is how we have seen Sam earlier.

Viewed later in their course, the maturing granuloma hangs from a thinner and thinner stalk before it falls off or is coughed off.

At times, when pedunculated on a long enough stalk, the granuloma flips up, out of the way during phonation and does not impair closure, so the vocal capabilities testing can return to normal while the granuloma is still in place, but on a long stalk. From injury, to ulcer, to granuloma, to falling off runs a course of a several weeks to several months.

Sam came back a year later.

"About two weeks ago, I was yelling at a baseball game. I coughed out something that looked like a BB and felt a sharp pain on the left side of my throat."

I took a look at Sam's vocal cords again. There is a granuloma on the right and a fresh wound on the left, likely the base of the stalk of the recently coughed out granuloma. Since this is an open wound or ulcer, he feels a sharp pain from it. As we know, this pain will often radiate to the ear because of neural pathways in the brain.

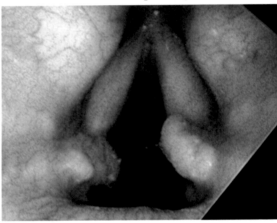

There is a healing stalk on the left side, just above another ulcer. There is a white granuloma on the right side.

If Sam's granulomas were from a previous intubation injury, they would likely not recur, because the intubation trauma is not recurring. The vocal process should have healed after self-decapitation of the granuloma.

130

His injury is more likely from compensation for a vocal weakness. My prediction is that unless Sam corrects the underlying bowing, these ulcers and granulomas will continue to recur. If he would stop speaking, they would also go away, though that is a bit of an impractical solution. If he cannot build up his vocal cord muscle mass through therapy and use, then one option would be to place implants into the vocal cords to artificially bulk them up. He would then no longer need to close the vocal cords as forcefully.

If the granuloma is removed surgically and the underlying cause is not addressed, surgery might be performed again and again, as each surgery opens a fresh wound at the base of the granuloma that can then be traumatized again, restarting the cycle of granuloma formation. At a minimum, voice rest after surgery until the mucosa has healed should reduce the frequency of recurrence.

I find other common scenarios to include granulomas in male executives in their fifties; people who feel compelled to speak with authority, but might actually have vocal cords that are starting to bow. I also see them in younger females who are trying to give an extra sense of authority by speaking with a lower pitch and more forceful voice. I have also seen them form from a chronic cough. Suppressing the cough allows the granuloma to mature, fall off and the vocal process to heal.

Granulomas are the condition first prominently conjectured to be associated with acid reflux[6]. They are often recurrent and form over the vocal cartilage within the larynx, rather than on the outside near the esophagus. The explanation invoking acid reflux fails to explain how acid would skip all the tissue between the vocal process and the esophagus and somehow always inflame the tissue over the vocal cartilage. External injury from intubation or internal injury

6 Koufman, JA. The otolaryngologic manifestations of gastroesophageal reflux disease (GERD): a clinical investigation of 225 patients using ambulatory 24-hour pH monitoring and an experimental investigation of the role of acid and pepsin in the development of laryngeal injury. Laryngoscope. 1991 Apr;101(4 Pt 2 Suppl 53):1-78.

from the vocal processes striking each other are explanations far more commensurate with the history and the visible findings of the disorder.

Muscular – Atrophy

Vocal underdoers – bowed cords – not talking enough

"I sound wheezy and hoarse," complains Joe Longevity. "For the past 10 to 15 years my voice has been cutting out on me."

"Anything else?" I ask.

"If I talk very long my voice squawks. Then it goes to a whisper for a few minutes. Sometimes it doesn't work at all."

"Is it effortful to speak?"

"Yes."

"Can people hear you in a restaurant?" I ask.

"No," he says.

"Anything else?"

"People either can't hear me on the phone or they think I am a woman."

"I see. What is your occupation?"

"I'm retired now."

"What did you used to do?" I continued to pry answers out of him.

"I worked on railroads on the cross country routes."

"Did you talk much on that job?" I query.

"No."

"Anything else?"

"I clear my throat frequently to try and get my voice to work," he replies.

"Have you seen anyone else about this?" I ask.

"A couple of years ago a doc looked at my vocal cords and said one of them might be paralyzed."

Joe rates himself a three on the seven-point talkativeness scale.

Joe has been a vocal underdoer for many years. Retirement has meant even less talking and now when he does socialize, his voice is effortful to use, it is undependable and his neck hurts afterwards.

Vocal underdoers suffer from deterioration of the thyroary-tenoid (TA) muscle. The surface mucosa is fine. The TA muscle responds to lack of use with atrophy and loss of muscle bulk so the term bowing is often used.

Ageing aggravates the problem. The elasticity of the tissues fades with age (think of the sagging facial skin of an elderly person) and in a recurring cycle, as the Joe's voice deteriorates, he withdraws from socializing. Withdrawal further diminishes vocal use leading to further thinning of the muscle. Since bowing often occurs in older individuals, it is also called presbyphonia (literally "aged voice").

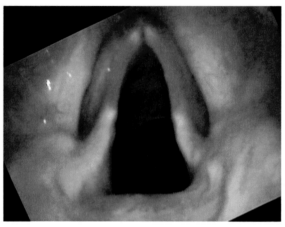

Muscle loss is three dimensional, so not only is there the appearance of bowing when viewed from above, but there is very little muscle mass within the vocal cords to even tighten during sound production. Often cords this thin will flutter during phonation.

With very thin cords, the vocal underdoer can place his vocal cords in a position for a low vocal pitch, but without rapid air expulsion through the large gap, the vocal cords fail to entrain. The CT muscle frequently contracts to stretch the vocal cords and place him in falsetto and is frequently used to compensate for a large gap. Joe is obliged to unconsciously use the cricothyroid muscle to make sound so we can say he has an *obligate falsetto*.

On vocal capabilities testing we find that the relaxed speaking pitch is higher than we would expect for Joe, nearing the female range. His lowest pitch is higher than normal. The uppermost note is normal. When I ask him if he can yell, he said that he hasn't tried. His yell is normal in terms of both pitch and volume. Often the patient has not tried yelling. Remember, vocal underdoers are likely naturally not very talkative.

When I got Joe to yell during the examination, his daughter was surprised at how normal his voice sounded. She cheered him on with, "Dad, that is how your voice used to sound!"

Now when we try the testing for vocal swellings at soft volumes, we will notice two findings. High in his pitch range, he is able to fairly easily make a soft sound without onset delays. However, as he progresses lower and lower he is forced to relax his CT muscle and the vocal cord bowing or gap gets larger and larger and at some point he is not be able to make a sound as the air just passes through the gap in the vocal cords without starting them vibrating. This onset delay is less precise than the findings with vocal swellings where there is a discrete note where the bumps touch. In this case the gap is progressively larger with lowering of the pitch, but Joe is variably compensating with other muscles each time I elicit the task, so the note at which there is loss of voice at low pitch varies a bit.

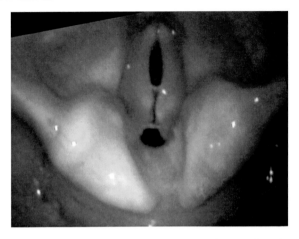

At a low pitch, when the vocal processes touch, the vocal cords are thin and bowed. Air leaks between his vocal cords and also behind the vocal process cartilages.

On endoscopic examination, two compensatory findings are common. If the person is not compensating (rare), then there will be a large central gap that will increase in width lower in the vocal range. However, with a long history of a soft voice, Joe's vocal cords cross each other posteriorly with the vocal processes overriding each other looking like scissors. This crossover reduces the size of the vocal gap.

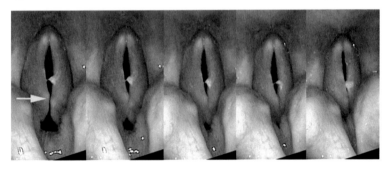

With effort, the vocal processes cross over one another in a scissoring fashion, pulling the vocal cords tight. Here the left vocal process (arrow) overrides the right in this photo sequence. This closure requires a lot of effort and also raises the speaking pitch.

Alternatively or additionally as the pitch goes lower, the person squeezes harder trying to approximate the vocal cords and visually the supraglottis (the part of the larynx above the vocal cords consisting mostly of the false vocal cords) squeezes together. In other people, the arytenoids squeeze closer and closer toward the epiglottis reducing the opening from front to back. Either way of squeezing, side to side or front to back, the supraglottis will progressively obscure more and more of the vocal cords, often completely hiding the vocal cords during the endoscopic exam.

Joe decided not to have anything done. He noted at the end of the visit that the problem wasn't bothering him enough to go through a surgery, but he was glad to have learned what was going on.

The second typical case below is in a young person who is innately not talkative. Think of the shy or introverted student in the class who never spoke to anyone and later makes a living working nights and does not need to interact with anyone. Her vocal cords thin prematurely.

The quiet person's job choice

It seems that vocal underdoers naturally seek out occupations that allow them to be quiet. I suppose it is the reward for matching occupation to personality. Someone whose batteries are drained by talking will be much more satisfied with a quiet occupation as typified by librarians or laboratory researchers where they can reduce unnecessary vocal interactions.

One day though, that librarian or researcher is promoted to team leader in management and now must conduct more frequent meetings. Her vocal cords are underdeveloped and now are called upon heavily. Like muscles elsewhere, if they are suddenly called into use, they fatigue easily, begin to ache with use and her voice fades out with use as the muscles give out.

In summary, if a patient who is naturally quiet or untalkative complains that they cannot be heard by others or they have laryngeal discomfort with required speaking, I will suspect and search for signs of reduced TA muscle mass. The vocal capabilities should worsen lower in the range as compensation to close the glottic gap is taken away.

Nonorganic

Improper technique

When there is no structural problem and "behavior" is the problem.

This vocal impairment results from vocal technique issues that start for a variety of reasons. I call this category *nonorganic* and place it in the behavioral hoarseness category because the mucosa is normal, the muscles are normal, all the structures are normal. The pattern of use of the vocal cords is the problem. This is the category where terms are most variable in my profession.

I find two subdivisions helpful. The first type of nonorganic patient is typically a singer, usually young, who complains they cannot sing as well as they used to. They may feel a lot of tension or discomfort. The second type of patient is one where there is some secondary gain perpetuating a voice problem that otherwise would resolve.

Muscle tension dysphonia

Competitive muscles

Maria Cantata is 18 and just started her professional voice studies at university. She began struggling with her voice and her new vocal professor suggested that she have her vocal cords examined before he pushes her too much. Maybe she has, God forbid, nodes![7] She has five years of formal vocal training and her goal is to become a profes-

---

7   It is my impression that classical singers have an almost religious fear of developing "nodes," as if the diagnosis of "nodes" represents everlasting loss of their career. Cognizant of this extreme anxiety, I spend quite a bit of time discussing their etiology and management. Even when I do not find them, I am sure to discuss their absence in a classical singer. I also prefer the term "vocal swelling" as it carries less baggage than the term "vocal nodes".

sional opera singer. During her last year of high school she sang with a pop band on some weekends. Now she is only pursuing classical musical training.

During our exam, I ask Maria to sing, going from a low note to her highest possible note. As she performs the task, ascending the scale, she sings louder and louder. Clearly she has a broad vocal range in terms of pitch. I ask her to repeat this vocal task again, but with the softest possible sound she can make for all the notes. She performs fine throughout her low range, but as she reaches her mid to upper range her voice starts cutting out and she becomes flustered.

"I haven't had a chance to warm up," she apologizes.

For diagnostic reasons, I am not interested in her *warmed up* voice, nor in her loud, robust voice. I ask her to start again at a slightly lower pitch and move up the scale, but she must not increase her volume. On a slightly different note, her voice begins to cut out.

What is happening?

I place my fingers on the top of her larynx as she sings. The muscles holding her larynx in place in the neck tighten. I feel her thyroid cartilage pull fairly strongly upwards in her neck during phonation. This excess tension causes the discomfort with vocal use, especially singing.

I record video of Maria's vocal cords at the pitches where she is struggling, noting that the back of the vocal cords are not together. Actually, when I slow the recording down, I see the vocal cords come together initially and then pull slightly apart as actual sound production begins. The LCA muscle brings her vocal cords together, but moments later she tightens her PCA muscle and this pulls the vocal cords slightly apart at the rear. At very slow speeds, I can show Maria this muscle bulging as it contracts.

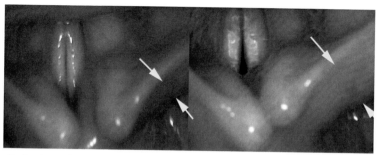

Left: *the right PCA muscle is relaxed as the vocal cords have closed. Between the arrows lies the bulk of the PCA muscle.*
Right: *moments later, the right PCA muscle has contracted (note the increased thickness of the right PCA muscle between the arrows) pulling the posterior vocal cords apart.*

When the vocal cords are held apart, air leaks between them. By asking Maria to sing at her softest pitch, the air preferentially leaks out through the gap, as a whisper, rather than starting up the vocal cord vibrations that we would hear as a note. The posterior gap between the vocal cords is variable depending on her effort and thus the vocal impairment is variable. That is why her voice starts cutting out on different pitches each time we test her singing up the scale.

Maria innately knows from practice that if she warms up, she can produce a sound from the very first attempt. Effectively after warming up, she builds up a muscle memory, a vocal cord position with a sufficient force of air beneath her vocal cords, that when released suddenly, as a blast of air, the vocal cords jump-start into vibrating. But she cannot get them started up with a soft, gentle onset. There is not enough air pressure to start the vibrations when the air can leak out through a gap more easily. Warming up in this case is the development of compensation for her problem. Since we are both in the office to find the problem with her voice, warming up is not helpful diagnostically.

In all likelihood, Maria trained herself to sing in a style with the vocal cords slightly apart, and that style was fine for her high school pop band. It just does not work for some of the classical operatic arias she is now attempting. Without going into all the inferences about what this means for training and how to correct the problem, we can at least identify the problem as an air leak. It has the quality of a husky hoarseness. If, at some time in the future, Maria learns to close her vocal cords completely while singing, her hoarseness or vocal impairment will resolve. Most likely this will require further work with her vocal coach. She will then be able to sing both softly and loudly at all the pitches available to her, within her vocal range.

Young singers often try to mimic a certain style or individual and in doing so often tighten the laryngeal muscles in an unusual way to achieve the sound they desire. One of the problems with mimicking something one hears is that one hears her own voice via sound conduction through the bones in her head, which is different than how others hear you. Thus, the reason no one ever likes hearing a recording of their own voice – it is *not them*.

Tension is the cause of Maria's air leak and discomfort. Relaxation of her larynx during singing will restore a normal voice. Usually directing a person's attention away from a particular sound quality and focusing on making each note clearly, no matter the character, yields the cleanest singing result. It can take time to correct this problem, as the muscle memory from singing a particular way for a prolonged period can be quite strong.

Nonorganic dysphonia

Psychogenic or secondary gain

The other type of nonorganic dysphonia or nonorganic hoarseness presents in many different ways. Really, a nonorganic hoarseness can mimic any organic voice disorder. Some of my colleagues term all of these disorders *Muscle tension dysphonia* though not all nonorganic dysphonias involve tension. Nonorganic hoarseness can also be caused by excessive relaxation of the vocal cord muscles.

While the problem in nonorganicity is in the brain, and not in the vocal cords (the organ), an alternative term "psychogenic hoarseness" carries with it pejorative meaning, at least in the mind of the typical American patient. Using the term psychogenic dysphonia can make resolution of the problem more difficult.

Additionally the American health insurance system often separates mental health care from "medical" care and will not pay for "psychogenic" disorders. Unfortunately, this pays for the physician not to diagnose a psychiatric condition.

With these caveats in mind, let's explore nonorganic voice disorders.

Andrew Smith first lost his voice three years ago. He is a professor and teaches at a community college. The first time he lost his voice, it was gone completely for a week: then he was hoarse for a few months. He saw an ENT physician who told him that his larynx was red and inflamed and placed him on a reflux medication – a strong proton pump inhibitor. This treatment helped "a little bit." However, when his voice returned, it was never as strong as it used to be.

The next school year he used a microphone to teach in class because his voice never completely recovered. Still, it was difficult for his students to hear him by the end of the day. By the end of each week, his voice was a whisper.

He improved during the summer. Then, the following fall, his microphone was not available and on the second day of teaching class, he lost his voice again, completely. He continued to teach by assigning projects rather than speaking about them. He cannot yell, primarily because it causes pain in his neck. A second ENT noted an ulcer on his vocal cords, treating him with antibiotics, steroids and an anti-reflux medication[8], ultimately for over six months and for most of that time he was on double doses. He altered his diet and his lifestyle to "avoid acid reflux" and he acknowledges, there was

8 Omeprazole (Prilosec™) is a proton-pump inhibitor and commonly used medication for the treatment of real or presumed "acid reflux."

perhaps slight improvement. He later tried a naturopathic approach without improvement.

As I try to reconstruct his story, his initial voice loss could fit the story of a person who develops a viral nerve injury. This acute laryngitis should have improved and not persisted. He is not an innately talkative person, so nodules are rather unlikely. One of the commonalities of nonorganic disorders is that they don't quite fit any specific organic illness set of symptoms. They come close, but there are discrepancies, so I begin to consider nonorganic hoarseness as a diagnosis.

The likely driving force behind a nonorganic hoarseness is that there is likely some secondary gain (obvious or not). The secondary gain may range from increased attention from family or friends, to relief from some obligation, to financial gain. Most times, the onset of the gain is initially inadvertent.

When I examine Andrew's vocal cords with the endoscope, they appear completely normal in structure. The visible functional abnormality is the vocal cords being held apart during phonation, though I could see them close together at various other times during the exam. Since they can close completely some of the time, I know they are not weak or paralyzed. I can feel with my fingers on his neck that he is pulling his larynx up very high and tight when speaking and also contracting the muscle under his chin very tightly. I find no red color out of the ordinary for mucosa. I find no ulcer. The only reason for air leak is the vocal cords being held apart during attempted voice production.

Now the story fits a pattern. Andrew likely had an initial organic illness, quite possibly acute viral laryngitis. With swollen vocal cords, in order to speak at all, Andrew had to hold the vocal cords slightly apart. However, as the swelling resolved and if, during the initial illness, Andrew was relieved from some burden, perhaps related to his teaching, then as the vocal cords returned to their normal size, the vocal cords tended to stay apart and his voice just tended to remain poorly functional. Holding the vocal cords apart now lets air leak out – a husky hoarseness.

One illness, viral laryngitis, was transmuted into muscle tension within the larynx, which holds the vocal cords apart whenever making a sound. It also explains the variety of symptoms within his illness. Sometimes he can only whisper, sometimes there is pain in his neck if he speaks louder than a whisper, sometimes his voice just "cracks," sometimes his voice returns completely to normal. He seems to "maybe get better" on anti-reflux medication or other treatments. Both his hoarseness and his improvement are inconsistent.

During my examination of Andrew, before the endoscopy, I record him performing a number of vocal tasks. I ask him to sing low notes and high notes. I ask him to shout, to cough, to clear his throat and to hum. Even though he is whispering to me at the beginning of the exam, on some of these vocal tasks a normal voice (perhaps inadvertently) comes out. During the endoscopy, I aim to reproduce these sounds which come out normally so I can film the position of the vocal cords.

At the end of endoscopy, I have a recording of both abnormal voice production and normal voice production, which aids me in diagnosis, but this is also the beginning of treatment. I review the video with Andrew and his spouse, who is present during the exam. I explain why sound comes out normally sometimes and why at other times, when the vocal cords are apart, sound is poor. Together we begin to practice the production of normal and abnormal voice. I explain how the pain comes from excessive muscle tension in his neck when speaking.

I persuade Andrew to buy in to his own ability to cure his voice. I am very successful in dealing with nonorganic hoarseness since the potential for complete return of normal voice is essentially 100 percent. Surgical voice problems do not have that rate of improvement since there is often some trade-off involved with surgeries. Andrew is re-learning to use his voice and by the end of the appointment we both have the visual and audible evidence that it is possible for him to restore his voice with practice.

Part of the treatment is having family present, so that when Andrew leaves he will have some support that his voice is recoverable. The family will have heard his normal voice. I write letters to all of the Andrew's other physicians to be sure every person treating Andrew is on the same page. Otherwise, it is often easier for a patient to remain hoarse and on a pill than to confront the issues and put in the effort to produce a normal voice.

His wife chimes in, "Does that mean we can lower the head of the bed now? Can we eat chocolate again? Can we eat Indian food? Can we have coffee in the morning again?"

Not only can Andrew resume his normal speaking pattern, his family can resume a normal life. I encourage Andrew to stop taking reflux medication, which is doing nothing for his voice, it may well have side effects and even without side effects it gives him a crutch to lean on, preventing recovery.

Andrew's secondary gain is subtle and not obvious to me. It has not been necessary for me to identify and correct the initiating problem. Just giving Andrew a face saving means of vocal restoration is often sufficient. I usually ask patients to return in a month, encouraging them to continue the practice we started in the office of switching between a normal and an abnormal voice – essentially giving the patient control over their voice again.

Many times when I see patients again, they cannot even make their "old, tight voice" anymore. They will report to me that suddenly one day their voice was normal again and now it is difficult to produce the tight voice without a great deal of effort and pain.

A resolution for nonorganic hoarseness can be made much more difficult when there is money, an outstanding workman's compensation or a legal battle involved. Especially difficult are cases where the patient will receive more money if they remain ill than if they recover. Until that financial carrot is removed, voice recovery is very difficult.

Another observation I suspect the patient cannot make in himself, but the examiner may observe, is an inappropriate smile which the examine can elicit; *la belle indifference* – the beautiful indifference. During the interview I often offer sympathy for a patient with a comment such as, "This vocal problem must be really getting you down." The nonorganic patient responds with a "Yes, it is really terrible!" accompanied with a rather bright smile on his face, the seeming opposite of the weariness one would expect with a drawn out, never ending, hopeless type of illness. An individual with a nonorganic illness is beautifully indifferent to his problem.

Nonorganic dyspnea

The voice is the seat of emotion

Julie Soprano arrives in the waiting area, breathing noisily. There is the temptation to pick up the phone and call 911 to ring an ambulance and have her taken back to the hospital. But after watching her for a few minutes, her breathing stays the same, a raspy noise with every breath in and out. She comes back to the examination room. She is able to speak, but her mother gives most of the history. Two weeks ago she developed trouble breathing and was admitted through the emergency room to the hospital in her hometown. She was breathing noisily then, and was given aggressive treatment for asthma, including nebulized epinephrine, steroids and other inhalers. She was admitted to the Intensive Care Unit for two days and was also given anti-anxiety medication. Despite this rather significant intervention, she failed to improve. After two weeks of hospitalization with minimal improvement, she was sent to the laryngologist.

The principle behind diagnosing laryngeal (and all voice) disorders, is knowing that each disease fits a pattern, with the exception of nonorganic illnesses. Diseases fit a pattern because each particular injury affects air movement through the vocal cords in the same way each time. For example, in a bilateral vocal cord paralysis, both cords tend to end up near each other in the middle of the larynx. This means there is very little space between them, whether for breathing or speaking. If you pass air between two vocal cords that are close together, they will vibrate. So a person with a bilateral paralysis will have noise when breathing in, noise when breathing out and they will create sound during speech that is often normal in volume. These symptoms are based on the physical location of the vocal cords in a fixed position near each other.

Comparing Julie's breathing to a patient with a bilateral paralysis reveals an incongruence for Julie. She is making noise breathing in, noise breathing out and yet she is having great difficulty speaking. Fortunately for me, the diagnostician, Julie doesn't know which symptoms go with which diseases. In nonorganic disorders, the symptoms are variable and inconsistent and I have already noted one inconsistency in Julie.

I turn on my recorder and listen to Julie speak. I ask her to perform a number of vocal tasks – reading, singing high and low notes, making a sound for as long as possible, yelling, coughing and throat clearing. Since each of these vocal tasks has a specific result based on how the vocal cords are positioned and how well they are functioning, I begin to hear the inconsistencies in her exam. Her symptoms are not matching up with asthma. They do not match bilateral vocal cord paralysis either. Both these diagnoses could be entertained based on some of her symptoms.

For instance, she coughs just fine. Coughing normally would be odd sounding if the vocal cords were fixed close enough together to cause noise with every breath in and every breath out.

I insert my endoscope so that I can record Julie's vocal folds. First, I observe breathing. Indeed, her vocal cords are very close together both on inspiration and on expiration. That tells me that a simple exam with an endoscope during her recent hospitalization could have alleviated all the "asthma" treatments, since asthma does not cause the vocal folds to come and rest together. The vocal cords are creating her wheezing sound while breathing.

Since I earlier heard a normal cough, I ask Julie to cough while recording the video. Julie's vocal folds open completely during the cough, then promptly come together again. With one vocal task, we have eliminated the possibility that the vocal folds are paralyzed. We have also eliminated the possibility that they are scarred together. We have documented the source of the sound – her vocal folds, not her lungs.

The treatment of Julie's problem is beginning, even in the middle of the examination. I continue to record the movement of the vocal cords under various tasks – attempted high pitch, attempted low pitch, sniffing, throat clearing, humming. I may even hold onto Julie's larynx with my finger and keep it in one position while she attempts various tasks, but above all, I am recording the exam.

Now that I know for certain that she has a safe airway (the vocal folds can open completely), I can ignore the noisy breathing as not dangerous. I play back the video to Julie and her mother, explaining how well the vocal cords open and close. I assert that although I don't know how the vocal cords came to be so close together, whatever got them to that position originally is now gone. We need only to retrain them to open with Julie's control in order to restore Julie's voice as well as her breathing. Interestingly she is a singer and actually had a competitive performance coming up later this week at school, from which she is presently excused because of her condition.

A nonorganic laryngeal disorder can mimic any other organic disorder, at least superficially. Typically a patient will not know all the symptoms for any given disease so the examination will have inconsistencies. One of the reasons nonorganic disorders so closely mimic organic disease is that frequently an organic disease starts the problem with the vocal cords. Then the secondary gain from being ill is so strong, that the patient avoids (consciously or unconsciously) getting better. It is more rewarding to stay ill than to get better.

I suspect that the relief Julie experienced from her initial illness (noisy breathing perhaps from a laryngitis that caused swelling of her larynx) included that she felt relieved that she would not have to compete at the upcoming vocal competition. She didn't want to get well too soon, or she would have to perform.

The diagnosis and treatment are highly intertwined now because, if I can relieve Julie's noisy breathing, my diagnosis is proven correct. To that end, I set Julie up to succeed.

I review my video findings with Julie and her family, explaining the great relief I have experienced since her vocal cords can open all the way. Julie is now "trapped" in front of her family into improving.

Secondary gain is a basic human tendency. Responding to a reward is normal, not malicious, and Julie will stop the abnormal behavior when she consciously recognizes the behavior and recognizes that the behavior is causing pain and worry in others.

I attribute her initial illness to a probable laryngitis (organic or non-psychological cause) so that Julie is able to "save face." She did not likely create the initial problem and having an explanation that she can provide to others to explain her illness allows for a comfortable way out.

I explain that this illness is a muscle problem. The muscles of the larynx have learned a specific way of tensioning the larynx, which may have been quite helpful when there was swelling present. However, now that the initial problem is resolved, the remaining problem is merely the muscle tightness.

If necessary, I reinsert the endoscope so that Julie can see her vocal cords moving and begin training. On whatever task that opens her vocal cords, I return to that task. I point out where Julie seems to have some control. In this case, coughing and sniffing open the cords. I have Julie perform those tasks while watching the video monitor. Since she has a difficult time with control, I emphasize first that she do the opposite of what we want. Squeeze the vocal folds even more tightly together, making the breathing sound even louder, followed by a sniff. She can then begin to feel the difference between very tight folds when making sound and the very open vocal folds when sniffing. I praise her for any progress she makes, so that there is at least some benefit to improving – my praise.

The noisy breathing begins to subside. The vocal folds gradually loosen and Julie's voice gradually improves. She spends about two hours with a speech therapist to further practice this tensioning and loosening. When I re-examine her the next morning, the noisy breathing is gone, the speaking voice is entirely normal and the vocal folds move completely normally on endoscopic examination.

There are two additional steps for me. I inform Julie and her family that there is a high probability that this condition will return at some point. I ask Julie to demonstrate the noisy breathing again and then stop it at will. She now has control over the problem. She should practice tightening and loosening the vocal folds, so should something inadvertently trigger the noisy breathing again, she will have a method for restoring her voice and breathing.

Additionally, I correspond with all of her other physicians (there are many; she had consults during her illness with allergists, pulmonologists, otolaryngologists and others). Everyone can learn from this experience of misdiagnosis. Also, should any one of her physicians see a similar illness occurring again, they will have a higher suspicion that nonorganic illness may be playing a role and not be distracted by the inconsistencies.

The difficulty with diagnosing nonorganic disease is that nonorganic laryngeal disorders mimic almost all the other laryngeal conditions. They can mimic spasms of the voice, weakness, double pitches, speaking in falsetto only (called puberophonia in males, although the condition occurs in females as well) or complete aphonia (no sound at all).

Although nonorganic conditions show up elsewhere in the body, there is probably some propensity for them to show up in the larynx. These conditions are the extensive modification of sound that is normally used to primarily convey emotion. So disorders that affect one's psyche may have a high propensity to affect one's voice.

Malingering

The secondary gain is obvious

While I give patients the benefit of the doubt that they would not be intentionally faking a voice disorder, there are times when after initially correcting a nonorganic voice disorder, and pointing out to the patient all her inconsistencies, the problem persists. The patient clearly has the ability to correct her voice problem and does not. Usually with inappropriate persistence of a condition, the secondary gain becomes obvious to the examiner. Perhaps the partner's additional attention that is present during the disorder, a monetary reward for remaining ill or some other gain misaligns with the patient's incentive to recover.

Not always, but often enough, a patient will have a history of multiple other, somewhat ill-defined disorders (often also nonorganic) in other parts of her body. One of the reasons I correspond with all physicians involved in the care of the nonorganic patient is to alert other caregivers, raising their suspicion about any medical problems that are inconsistent with typical illnesses. Otherwise, patients with nonorganic conditions tend to gravitate to the physician who gives the most treatment. Not infrequently, since the patient is not responding in the usual manner, the physician administers stronger and stronger treatments and the patient begins experiencing actual organic side-effects from her medical or surgical treatments.

Errors in diagnosis

A caution

I have to be careful with this diagnosis of nonorganicity as physicians have a track record in the past of calling something they did not understand a psychogenic disorder. As new medical understanding comes along, an organic disorder may be discovered that accounts for the patient's unusual problem. For example, prior to the 1980s, patients with vocal spasms interrupting their voice were sent to psychiatrists. Interestingly, most or all failed to improve with psychiatric treatment. With the advent of an induced nerve paralysis by Herbert Dedo, MD in one individual, patients with this condition, later called spasmodic dysphonia (See "Laryngeal dystonia" on page 206), found an effective treatment for their neurologic condition.

Structural hoarseness

Structural disorders

A change in anatomy changes the voice

 There are several categories of vocal problems where something other than vocal behavior changes the structure of the vocal cord:

- Infectious disorders

- Congenital disorders

- Neurologic disorders

- Tumors

- Mucous problems

- Traumatic disorders and

- Unusual disorders.

Infectious disorders

Acute viral laryngitis

When you catch a virus in your larynx

I felt a tickle in my throat around 6 PM. By the time I went to bed, there was a noticeable discomfort in my throat. At first I had to clear my throat and later I began to cough. The following morning my throat was sore and my voice was deep. I had developed an acute laryngitis. I was heading into the office anyway, so I took the opportunity to have a look with my endoscope.

My vocal cords were noticeably thicker than normal. On endoscopy, they have a dull, swollen, glassy appearance. Secretions were unusually thick and sticky. I could sing lower notes than normal by more than half an octave. I also could not sing any upper notes at all. If the vocal cords swell enough, they may become stiff and unresponsive, but this time I did not lose my voice completely. Acute laryngitis is typically from a viral infection that causes the blood vessels within the vocal cords to dilate. Fluid leaks out from the blood vessels and accumulates beneath the surface of the vocal cords causing them to swell.

In this case, the usual treatment of fluids and time gradually allowed my vocal cords to recover from this presumed viral infection. Within three weeks my voice was back to normal.

I realize that it is seldom that a patient finds himself in a laryngology office within 12 hours of the onset of acute laryngitis for a visual endoscopic examination of their vocal cords unless he owns one. Unless breathing is impaired, it is unlikely that there is much to do about acutely swollen vocal cords other than treat the symptoms. If an acute laryngitis fails to significantly begin to improve within about two weeks, it may then be worthwhile seeing a laryngologist.

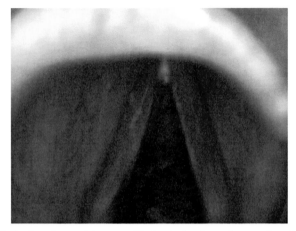

Acute infectious laryngitis, most likely viral, where the vocal cord's blood vessels are dilated to the point that the true vocal cords have almost the same color as the false vocal cords. Secretions are white and sticky.

Papilloma virus

Human papilloma virus

Darcy Loma noted her voice changing. Her primary care physician thought she might have a virus and treated her symptomatically. Her voice worsened and at an ENT physician's office she was told she had a polyp. She became pregnant about the same time as the onset of the hoarseness and at seven months pregnant she was now winded climbing a flight of stairs. She is snoring at night, especially if laying on her right side. Two weeks ago she "coughed up a piece of the polyp" and has since had the taste of blood in her mouth at times.

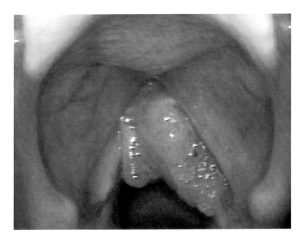

An irregular lumpy mass on the right vocal cord which is papilloma.

I biopsied this in the office, to avoid using a general anesthetic in a pregnant female and confirmed it was papilloma. We then decided to wait until the end of her pregnancy to treat it. By the end of her pregnancy she was a little more short of breath and her voice was completely a whisper.

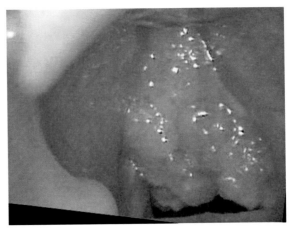

Three months later the papilloma had grown. Only the back end of the left vocal cord is barely visible.

While this is an infection, the papilloma virus causes a growth, typically with little or no inflammation. Since it behaves much like a tumor, it ends up being treated in a fashion similar to tumors – excision. Papilloma most commonly grow on the hands, the feet, the nose and genitalia. While it may be transmitted sexually on genitalia, no one yet knows how it ends up on the vocal cords.

Although we don't have great options for treating viruses medically, various antiviral medications are being assessed. Also, the development of a vaccine (Gardasil™ – Human Papillomavirus Quadrivalent [Types 6, 11, 16, and 18]) against the most common type of papilloma virus found on the vocal cords shows promise for preventing the disease on the vocal cords. Recent evidence may even be suggesting that the vaccine given after acquiring the virus might diminish the growth of the papilloma virus.

I excised visible papilloma from Darcy's vocal cords three times. I use a CO_2 laser to cut around the edge of the papilloma and peel it off the surface of the vocal cords. It does not grow down into the vocal cords and with this technique I can best preserve the vibrations and functioning of the voice.

There is a generally acknowledged high rate of recurrence of papilloma after excision and consequently there are many methods attempted to treat it. Various antiviral drugs have been injected or administered. Other surgeons, although they use a laser, may burn off or vaporize the papilloma rather than cut around it or they may use a device that sucks and shaves the papilloma off. I have some misgivings that these techniques can as completely remove papilloma while preserving as much of the underlying structure.

She has now gone eight years without a recurrence.

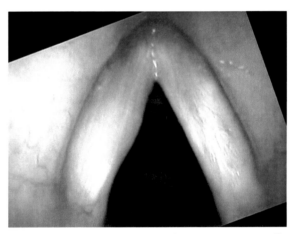

Eight years after her third excision of papilloma. She is very talkative and has a callus on the left vocal cord. The right vocal cord has some scarring from the papilloma removal, but vibrates reasonably well.

With new high definition endoscopes[9] and video processors that selectively highlight the red in blood vessels, papilloma can now be fairly accurately identified endoscopically in the office.

---

9   I use the high-definition Pentax VNL-1590STi endoscope coupled to a Pentax EPKi video processor to obtain 720p HD images. The color altering feature is called iScan and has a number of settings. Olympus began making a high-definition laryngeal endoscope in late 2011 which also alters the colors to emphasize certain lesions which also uses a technology they previously developed called narrow band imaging (NBI).

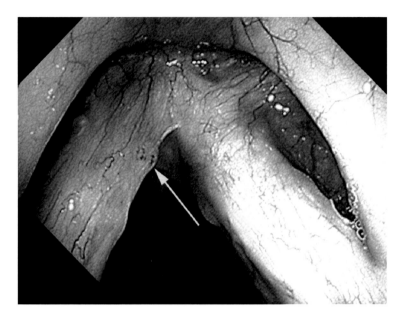

High-definition endoscopic view of recurrent laryngeal papilloma in a different patient, viewed with artificial color to emphasize blood vessels. The papilloma cells have a small vessel loop in each one of them, giving them a red-dotted appearance. The irregularity of the other blood vessels in this patient is from prior excisions of papilloma. When new vessels grow into the vocal cords after surgery, they follow an irregular pattern.

Small amounts of papilloma can also be treated in the office by ablation with a laser such as the pulsed KTP laser which may be passed through a channel in the endoscope. The energy from the pulsed KTP laser is selectively absorbed by blood vessels and since all papilloma have a large central vessel they are particularly susceptible to destruction by this laser.

Fungal laryngitis

Steroid sprays – what you don't see, matters

Barbara S. Ma has been hoarse for several months. She appears a bit older than her age, with the fine facial wrinkles of someone who has smoked tobacco for some time. She attributes her current problem to her husband, who several months ago began smoking again. Although he had stopped for a year, he resumed smoking after a stressful occasion at work. Barbara is a former smoker and says that her hoarseness came on about a week after her husband resumed smoking. She is concerned that she may be developing cancer, both because of her former smoking and her husband's current second-hand smoke. Even though developing cancer a week after exposure to second-hand smoke doesn't sound credible, it is an issue for me to address as a physician as it is a real fear for Barbara.

Barbara clearly sounds hoarse as I interview her and I note that she is on a number of pills and medications. When I review her various medical diagnoses, I note that she checked off the asthma box, yet she is not on any medication for it. I inquire about her asthma and she says, "Sorry, I forgot, I am also on an inhaler. I am using Advair™ (fluticasone/salmeterol), but," she adds, "my asthma is under excellent control". She had been switched to Advair about 4 months before her hoarseness started.

When I place the endoscope through her nose and begin to view her larynx, the larynx appears with reddish hue. This is the view obtained by the typical otolaryngologist who examines the vocal cords from five or more centimeters away. The examiner doesn't want to gag the patient and the patient doesn't want to be gagged by the scope either, so the examiner looks at the larynx from above the epiglottis – too far away to see detail. Many physicians tend to focus on the back of the larynx and state that the "interarytenoid area is red, so it must be reflux."

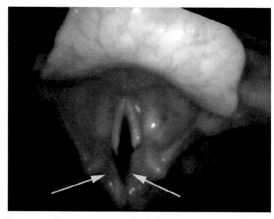

The area between the arrows is called the interarytenoid area and since much of the translucent mucosa is in a plane nearly parallel to the endoscope, the capillaries are seen stacked on top of each other giving the angled surface a darker red appearance.

Since I have never seen reflux laryngitis, I move the endoscope extremely close to her vocal cords. Sometimes this maneuver can be performed without topical anesthesia if done with caution, moving the endoscope gently forward as she breathes in and the vocal cords open widely and moving slightly away as she breathes out and the vocal cords come closer together. If unsuccessful in getting close, I take the additional time to drip a numbing medication on to the vocal cords, so that even if I touch them, there is no gagging.

As I approach her vocal cords, and ultimately I will be within one millimeter of the cords, there is no reddish hue. Redness turns into individual dilated blood vessels on the surface of the vocal cords.

From a distance, there are always a couple of illusions present. The throat is red because the mucosa covering the throat is translucent. You can see through mucosa. The red or pink color comes from the vessels beneath the mucosa. Moving closer, when mucosa is seen straight on, it is only lightly pink as the blood vessels are spread out. Yet when viewed in a plane nearly parallel to the endoscope, mucosa appears redder, because while you still see through the transparent mucosa, in an angled or parallel view, all of the vessels end up being stacked on one another, so visually there is more apparent redness.

Second, the further away the endoscope is from the mucosa, the more light that is required to illuminate the throat. Since the endoscope has a limited supply of light, the manufacturer of the video equipment uses the video processor to increase the gain or amplify the video signal. Gain allows the brightness to appear the same on a monitor whether close up or far away from the vocal cords. However, when the picture is made bright by increasing the gain, there is much more artifact or "noise" present. Video noise consists of more random spots, giving the picture a graininess. Many of these spots are red and consequently, distant views give the examiner the impression of more redness.

As I move closer, the video gain decreases and the picture clears. I start to make out the individual blood vessels that are present under the mucosa of the vocal cords. The vessels are a bit dilated or larger than usual, and yet we still have not seen Ms. Ma's problem. We are looking, but not seeing.

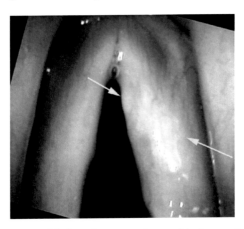

The area between and below the arrows has a thin layer of white fungus partially obscuring the blood vessels beneath. The film appears whiter on the edge of the vocal cord as we are looking through more fungus on the edge.

If we stop and search this close up view for a time, we will begin to perceive that all of the vessels dim in the central portion of the vocal cord. They are actually diving under a thin film of white. Because the vocal cord itself is white, a white film on the surface (white on white) is not easily discernible. But if we concentrate on the vessels, we can see a distinct curving line where the vessels fade or disappear as they traverse the central portion of the vocal cord.

I make a diagnosis of fungal laryngitis since I perceive a whiteness and the typical throat fungus is Candida albicans, which is white. Fungus growing on the vocal cords would also stimulate increased blood flow or inflammation. The patient is on a steroid spray that, while very good for the prevention of asthma, will encourage the overgrowth of the normal fungal flora in the mouth and throat (also known as thrush). I treat her with four weeks of fluconazole and teach her how to rinse her vocal cords off after each use of the Advair inhaler.

Typically when one gargles, all the water is in the back of the mouth since we naturally try to keep water away from the vocal cords. To bypass this tendency, a patient places a curved cannula in the back of their throat and while saying "eeeeeee" drips water directly onto the vocal cords. Because air is coming up while saying a sound, the water bubbles on the surface of the vocal cords and a *laryngeal gargle* is created. Yes, this can cause gagging and many people cannot do it. Yet, water gargled on the vocal cords can very effectively remove the inhaled steroid that has landed on the vocal cords.

Additionally, we ask if her lung doctor will allow a lower dose or a less potent steroid inhaler, reducing that treatment to the minimum required to treat her asthma.

At her follow up visit, one month later, she says that her voice started improving about ten days after starting the treatment and was completely normal by about three weeks. On her checkup, when I maneuver the endoscope close to the cords the blood vessels no longer disappear under a layer of white.

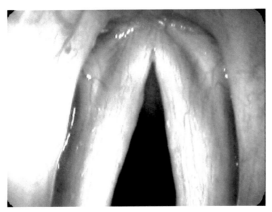

Vessels are now visible along the length of the vocal cords after treatment for fungal laryngitis.

I have never biopsied a vocal cord to look for fungus, but as the management of asthma has relied more and more heavily on potent steroid inhalers (and to some degree this can happen with nasal steroid inhalers or with other medications that cause immunosuppression), this finding of a thin white film and dilated blood vessels has become more common. Since all of the patients have responded to anti-fungal treatment, I have assumed the white film is fungus and patients have been rewarded by successful restoration of a clear voice.

When comparing the before and after stroboscopy views, one can also appreciate the stiffness of these lightly white vocal cords that improves after treatment. Barbara also did not have the same amount of fungus growing on each vocal cord, so the mass and the stiffness was different between the vocal cords and she had a very rough voice before treatment because each cord was trying to oscillate at a different pitch. It is also possible to have an almost symmetric stiffening or increase in mass. The voice will be smooth, but the patient will sense an increased effort with sound production, since stiffer vocal cords require more air to make them vibrate.

This frequent finding associated with the potent steroids being used to treat airway disorders has convinced me again that redness does not vibrate and degrees of redness, while partially from increased blood flow, are mostly an artifact of the camera system and video processing. The goal is not to see a color, rather to see what is stiffening the vocal cords, in this case the film of white fungus and the surrounding edema from a surface infection that causes stiffening of the vocal cords. After treatment, her vocal cords became supple again in a reasonable period of time – a few weeks.

I reassured Mrs. Ma that although her husband's smoking was not healthy, it was not the cause of her hoarseness and she did not have a vocal cord cancer.

I utilize her video recording frequently in my lectures to alert physicians that they may be taking pictures of the problem and looking at the problem without ever really seeing the problem. I also use them to emphasize the need for a very close, detailed examination.

There are more blatant examples of fungal infections. A suppressed immune system from treatments such as chemotherapy or systemic steroids for a prolonged period can lead to bright white, fuzzy patches growing all over the larynx and even the rest of the throat. These more extensive fungal infections appear the same on the larynx as oral thrush appears in the mouth.

I have also seen a few patients on steroid inhalers where the layers of fungus build up to the point that the lesion resembles a cancer.

Within three to four weeks of treatment with an antifungal the lesion disappears.

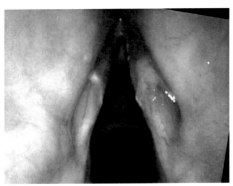

This right-sided, red, raised lesion with a central white area was from nasal steroid use and resolved completely within four weeks of antifungal treatment. This thickening is so extensive that it can resemble the appearance of vocal cord cancer.

Bacterial laryngitis

Acute infections – epiglottitis

I tend not to see bacterial infections of the larynx. I suspect that they are either self-limited and resolve from the body's own immunity in fighting the infection or quite possibly they are treated with antibiotics in the primary care setting.

One type of infection of the larynx often called epiglottitis, that perhaps more appropriately might be called supraglottitis, is typically a bacterial infection of the upper portion of the larynx. The vocal cords have a very limited ability to swell acutely. When they do swell even a small amount, they become rather stiff and unable to vibrate. The person loses their voice, but continues to breathe rather normally.

When a bacterial infection involves the upper part of the larynx, the supraglottis, a great deal of swelling may occur and it may progress rather rapidly over hours. The mucosa or surface tissue is very loosely attached especially at the back and a great deal of fluid can accumulate beneath the tissue. So much swelling may occur that the tissue begins to suck into the airway creating noise and difficulty breathing. The epiglottis is often swollen, but typically it is the mucosa covering the arytenoids at the back of the larynx that is so loosely attached that when swollen, the supraglottic mucosa begins to suck into the larynx when breathing in. This type of infection has been known to completely block off the airway.

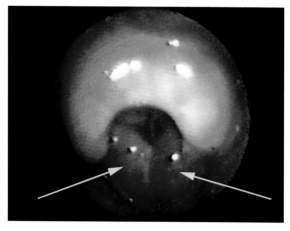

While the epiglottis is swollen at the top of the picture, it is the swelling of the arytenoids (arrows) that begin to pinch off the airway or suck into the opening between the vocal cords when breathing in, blocking off breathing.

Occasionally I see a low grade smoldering infection of the larynx that presents as apparent dryness of the vocal cords along with some dilation of the capillaries on the vocal cords with stiffness. While this rather benign dryness doesn't behave aggressively, a culture taken from the surface of the vocal cords can reveal bacteria and treatment can then be based on the bacteria found.

Unusual infections

Coccidiomycosis – Tuberculosis

Jesus Juarez had difficulty breathing starting three months ago. He thought it was just a cold but it progressed. In the emergency room he was given epinephrine and steroids to help his breathing. When asked about prior problems, he recalls that 20 years ago he was hoarse for a year while living in central California. He had nodules on his vocal cords that when biopsied, he was told that he had "Valley Fever" or coccidiomycosis.

On his close endoscopic examination there was a bulbous collection of giant cells growing out of the left arytenoid area. When something with an unusual shape or form is growing on the larynx, a biopsy is a very appropriate next step. The office biopsy revealed coccidiomycosis had invaded his larynx and was preventing his vocal cords from opening well.

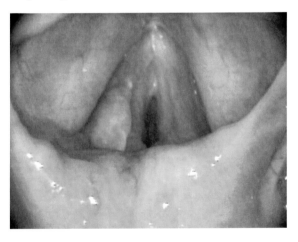

Coccidiomycosis growing on the back of the left vocal cord and into the left arytenoid.

Moon Lee moved to the United States five years ago. Last autumn, he returned to Korea for a conference and developed a cough that persisted on his return. By December, he had lost much of his upper singing voice and was persistently hoarse even when speaking. He could not project his voice anymore. An ENT told him that he had reflux laryngitis and recommended that he avoid spicy foods.

On my endoscopic examination, there were irregular swellings on the edge of the left vocal cord. This was my first view of tuberculosis on the vocal cords. There were cells of *Mycobacterium tuberculosis* in his sputum when cultured. He was referred to an infectious disease colleague and received extended treatment for tuberculosis.

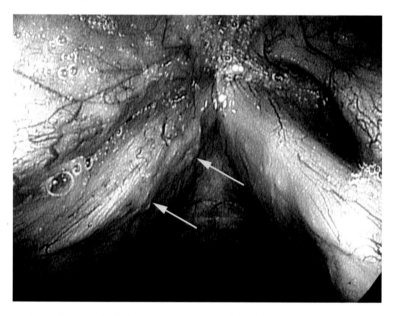

Tuberculosis on the left vocal cord margin (arrows).

Congenital

Sulcus vocalis

Congenital missing muscle

Pandit, a young man in his thirties, called me asking if I could lower the pitch of his voice. He has been called "ma'am" on the phone far too often. His voice has been high pitched his whole life. The louder he tries to speak, the more his pitch will go up. He frequently "squeaks" when speaking.

On endoscopic exam both vocal cords are bowed and very thin. There appears to be a groove running along the length of the entire vocal cord. That groove could be a scar, but seems to me to represent such significant atrophy of his vocal cord muscle or TA muscle, that the vocal ligament is the main supporting structure of the vocal cord. The mucosa appears to be draped over the ligament, then there is the groove and below that, perhaps a tiny bit of muscle.

You might reasonably ask how a person can speak without the main muscle inside their vocal cords. Effectively Pandit is speaking in falsetto full time. Because we have two muscles to raise pitch (TA and CT), if one is missing or impaired (the TA muscle in this case), the other muscle takes over as compensation. The CT muscle contracts, pulling the vocal cords longer and tighter. This brings the bowed vocal cords closer together so they can vibrate more easily, but it also raises the pitch.

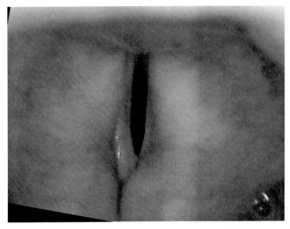

Rigid endoscope view of closure. Despite squeezing of false cords, there is a large gap between the cords.

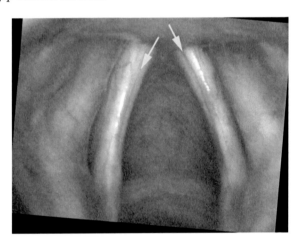

Arrows point to grooves in the very atrophic vocal cords.

It seems to me, the most likely etiology for this disorder is that a person is born with small thyroarytenoid muscles (or very atrophic muscles or with a damaged nerve to the muscles – they are innervated by the last branch of the recurrent laryngeal nerve). It is certainly plausible that some infection causes a destruction of the tissue within the edge of the vocal cords.

The best results I have had in accommodating this problem are to place bulking implants in both vocal cords. That adds only a little tension, but the bulk places the vocal cords closer together and increases their mass. The cricothyroid muscle then contracts less, so the relaxed speaking pitch may drop.

Transgenderism

Born into the wrong body

Once going by the name of Michael Strong, she is now known as Michelle. Michelle wishes that her voice more nearly matched her appearance. She has spent several years of her life and many thousands of dollars changing her male birth anatomy to the anatomy she identifies with, which is female. She has had genital gender reassignment surgery. She takes female hormones. Her breasts have enlarged and she helped them further with implants. She dresses impeccably. Her facial bones have been re-contoured. However, on the telephone, she is frequently called "sir."

I also meet transgender males who were originally females. As a broad generalization, since the genetic female's larynx still approximates a young boy's larynx in size, cartilage softness and consequently in pitch, with masculinizing hormones (testosterone) the pitch will generally gradually drop. Effectively this new male will now go through puberty with their voice over somewhere from one month to two years of hormonal treatment.

In the other direction, a genetic male such as Michelle, who is now living in or transitioning to a female life, has the unfortunate problem of the larynx already having enlarged to create a tenor, a baritone or bass voice. On top of that, testosterone promotes calcification and hardening of the laryngeal cartilage so that it becomes bone. It will not shrink no matter how much estrogen is added to her body. Consequently, males transitioning to females generally require some surgical change to their larynx if they do not wish to or cannot full time stretch their vocal cords into falsetto when making sound. It is possible to keep the vocal cords on a stretch full time, but very difficult. On top of that, the throat and chambers above the larynx are large and resonate and amplify the deeper pitches. In addition to tightening the cords, the transgender female must also learn to constrict her throat full time when speaking. All this is on top of learning the societal speech patterns of females.

I have been pursuing a surgery, feminization laryngoplasty (FemLar) for a number of years trying to address as many of these parameters as possible. Through an incision in the neck, I shorten the length of the vocal cords, try to tighten them so that at a longer length they are effectively a bit thinner. I reduce the size of the thyroid cartilage so that it protrudes less as an Adam's apple. I raise the larynx in the neck so that the resonating pharynx is shorter and amplifies more of the higher frequencies.

On endoscopic examination before surgery, Michelle's vocal cords would be considered normal male vocal cords. She underwent the feminization laryngoplasty surgery and has been living contentedly as a female since. Her voice continued to change for about one year after the surgery. Some of this may have been part of the healing process including gradual reduction of swelling and a more gradual softening of scar tissue. However, I think much of the change was Michelle's gradual learning or adaptation to the new amount of closure required of her shorter vocal cords.

There are several different approaches taken to raise the pitch of the speaking voice including:

- voice therapy alone,

- feminization laryngoplasty mentioned above,

- sewing the front of the vocal cords together,

- lasering the vocal cords or

- cricothyroid approximation.

Some people respond well to coaching and training. With therapy and practice they can produce a feminine sounding voice. Is that some people will never be able to sing well, some transgender females will never be able to produce a feminine sounding voice no matter how much they try.

Vocal cord webbing is a surgery done through the mouth where the front half of the vocal cords are sewn together. Lasers have been used to try and thin out the muscle of the vocal cord and to try and tighten the surface of the vocal cord.

Cricothyroid approximation (CTA) surgery is a simple approach that effectively tightens the CT muscle which raises the voice into falsetto, potentially giving the patient a permanent falsetto voice.

Unfortunately about one-third of patients after CTA surgery have their vocal cords stretch back out and their pitch drops back down to its previous range after a few months. It is also in the mind of the listener to judge whether a falsetto voice sounds more feminine or more like a gay male's voice. I dislike the quality, which coupled with the one third failure rate and an uncommon complication of locking the patient into a single pitch, led me to stop performing CTA surgery.

Neurolaryngology

Paresis – Recurrent laryngeal nerve

Injury and recovery of the main nerve to the larynx

Five years ago, at age 75, Peter McDash was found to have a thyroid cancer. His voice was normal at that time. He underwent radiation therapy and then surgery to treat the cancer and his voice was still normal immediately after surgery. Twenty-four hours later his voice became weak, higher pitched and he lost his vocal durability. Since then, he cannot compete with any background noise, such as in a restaurant, and that is a significant problem for a talkative person in a talkative family.

He initially sought help from several doctors. One doctor told him that one of his vocal cords was not moving but he should live with it. Another doctor told him that perhaps his thick secretions were straining his vocal cords. He became skeptical that anything could be done to restore his voice, so he has just got along since then without any treatment. Peter's son-in-law previously had a successful voice surgery and finally persuaded Peter to come for a visit.

On first listen, Peter sounds like a frail, 80 year old man with his voice thin and high in pitch, yet he plays tennis every day, suggesting he may not be as frail as he sounds. Thus, there is an incongruity in Peter's story; he sounds frail, but he is not truly frail.

After beginning the endoscopic exam, my first impression is that Peter's right vocal cord is in a position near the midline and not moving, while the left vocal cord opens and closes. Most physicians would agree that he has a paralyzed vocal cord. That may even be exactly what his first doctor said.

Merriam-Webster defines *paralysis* in two different ways:

- complete or partial loss of function or

- loss of the ability to move.

Peter does have a paralysis by this definition. One vocal cord is not moving open and closed when he breathes or speaks, while the other cord retains its opening and closing motions. The injured vocal cord appears (superficially) to be fixed in position.

As a physician scientist, I am ultimately interested in why something is not working because when I understand the why, I can develop a model for a disease and arrive at the most appropriate solution to a problem.

I ask, why did he lose his voice several days after surgery? Why is he hoarse five years later? Why doesn't he and his previous doctors realize that several other symptoms (that he isn't even telling me about and I will learn about later) are also related to the same injury? If we understand the mechanics of the normal vocal cord, why he has lost visible motion and why his voice is poor then we can answer these questions and we can offer solutions. These are questions that his prior physicians didn't completely answer. We can pose more questions and they should still fit the same explanation. For instance, why is his speaking pitch high?

Why did he lose his voice?

Let's start with the first question. Why did Peter lose his voice several days after surgery?

The main nerve that supplies the majority of the muscles of the voice box is called the recurrent laryngeal nerve (RLN). Due to something that happens when we are still not much larger than a ping-pong ball in a uterus, both recurrent laryngeal nerves (left and right) leave the brain, head to the chest and pass around a blood vessel before coming back up to the larynx. On the right side, this U-turn takes place under the collar bone, but on the left side, the U-turn is at the level of the heart.

These long RLNs pass a lot of important structures on their way to the vocal cords: the carotid artery, the windpipe (trachea), the swallowing tube (esophagus), the neck vertebrae, the lungs, the heart and the thyroid gland. If a tumor grows from one of these organs or if a surgeon works on one of these structures, there is the potential for injury to the RLN by stretching or by injuring the nerve's blood supply. If you pull on the cable to your telephone hard enough, you may lose the connection inside the cable. Even when the surgeon just pulls the RLN aside to get where he is going, the RLN can be injured to the point where less signal passes through the nerve.

An interruption of a signal going to a muscle results in a re-duced movement of the muscle supplied by that nerve. No signal to a muscle means no movement in that muscle. Superficially that would seem to account for Peter's condition. He has no motion on the side of his nerve injury.

But the answer is more complex. Each RLN, on each side, sup-plies several muscles including muscles that open as well as muscles that close the vocal cord. When the RLN is injured, the movements required for breathing (abduction) and for making sound (adduc-tion) are often injured simultaneously.

To think further about this let's ask, why didn't Peter get hoarse until several days after surgery? In a long surgery, under general anes-thesia, there may be a tube in the windpipe for several hours. It can cause swelling of the vocal folds, so both vocal cords may be swollen enough that with one RLN injured, the vocal fold still working can close against the swollen, immobile vocal cord and make a reasonable sound. As the swelling subsides, the opening between the vocal cords becomes larger and Peter loses his voice because he leaks air through this enlarging gap in the vocal cords.

Why did his voice come back?

Now the double question. Why did Peter's voice come back, yet this recovery was incomplete? Why does he remain hoarse 5 years later?

If the nerve injury is very mild, a reconnection may occur within a few days or a few weeks. The reconnections of the nerves will be appropriate in quantity and the nerve fibers will reconnect with their original muscles. This will get the vocal cord muscles contracting appropriately again. The voice is restored to normal and visible movement is restored to normal.

In anything more than a mild injury, which is the case with Peter, the nerve fibers regrow back through the injured nerve trunk over three to four months and reconnect to the vocal cord muscles. While the nerve fibers have an incredibly strong propensity to find a muscle to connect to, they are indiscriminate or not very choosy. The fibers will hook up to any muscle they encounter that needs a connection. A lot of the connections in the larynx become mixed up because all of the injured muscles are releasing nerve attractants. Some of the fibers programmed in the brain to open the vocal cords inadvertently connect to the muscle that tightens (TA) or to the muscle that closes (LCA) the vocal cord. Some of the fibers meant to close the vocal cord find their way to the muscle that opens (PCA) the vocal cord. With this random reconnection, all of the muscles initially suppled by the RLN may be reinnervated in new proportions or new degrees after the nerve reconnections are complete. Each muscle may receive appropriate as well as conflicting or inappropriate stimulation when the individual wants to make a sound, when he wants to breathe or when he wants to swallow.

For example, the LCA muscle is positioned to move the vocal cord to the midline, such as for making a sound. Let's say that the LCA now receives a signal during breathing in (inspiration) as well as a signal when we try to make a sound. Now it is stimulated all of the time, under all conditions (breathing, phonating, swallowing) and thus essentially remains in a continuous state of contraction. The muscle for opening the vocal cord, the PCA may also receive a constant signal now (during both breathing and trying to make a sound). The vocal cord ends up moving to the middle of the road position with all this stimulation and just stays put. The vocal cord is visibly paralyzed. It doesn't move, but it is not without innervation. The innervation is just continuous rather than intermittently appropriate.

The PCA muscle is the larger muscle, so if enough of these fibers end up going to the LCA muscle instead, the vocal cord ends up in a nearly fixed position somewhere near the midline. Quite often this frozen position is close enough to the midline that the remaining good vocal fold can get close enough during attempted voice production to set the vocal cords vibrating and create a voice. The good vocal cord can also open wide enough for the individual to breathe in comfortably even though the injured side does not open.

Shortly after an injury, the vocal cord is essentially paralyzed or not moving because it has no signal. Later, it has plenty of signal, but the competing signals counteract or balance each other, preventing effective motion and the vocal cord remains in a nearly fixed position. (This perfect balance only happens when the competing signals are equivalent. Quite often they are not in perfect balance and this will account for the wide variety of symptoms that can be attributed to a neurologic vocal cord injury. I term these disorders "dyskinesis" or dysfunctional motion, rather than "paralysis" or lack of motion.)

Several months after his injury, Peter's reinnervation fixed his right vocal cord near the midline and his left vocal cord learned to actually cross the midline. Peter's vocal cords could again nearly close, if not quite as completely as before his injury. He could set them vibrating and his voice returned. Let's explore some of the details for why his voice remains hoarse despite this closure.

You may remember from the chapter "Two types of hoarseness" on page 55, my proposal that hoarseness is only from an air leak or an asymmetric vibration. Lack of vocal cord movement (abduction or adduction) is not a cause of hoarseness. If the vocal cords are fixed next to each other and air is blown through them, they will vibrate and produce sound even though they do not open. We don't need to describe Peter's lack of vocal cord movement during breathing to explain his hoarseness, we need to find an air leak or an asymmetric vibration.

I put the scope into Peter, viewing his larynx, noting the obvious lack of opening and closing motion on his right side. But that does not mean there is an air leak. I search for an air leak by having Peter make a sound several times and then I carefully review this process of closing the vocal cords.

Watching the video in slow motion, initially I see his good side coming completely to the paralyzed side. Then as he starts to build up air pressure beneath the vocal folds, the paralyzed side is weak enough that it actually gets blown slightly sideways, away from the strong left vocal cord during the creation of the sound. A gap opens at the back of the vocal folds at the vocal process. Air leaks along the entire length of the cord.

This air leak affects vocal quality, adding a huskiness to his voice. Additionally, he uses up more air to make each sound, so he gets out of breath quicker with talking than he should. Now we have part of the explanation for the quality of Peter's voice and for his shortness of breath.

Unknowingly, Peter tries to compensate for this leak. The CT muscle that stretches our vocal cord and puts him into falsetto, is supplied by a different nerve (the superior laryngeal nerve or SLN), so is probably uninjured from the low neck operation. Peter's CT muscles stretch his vocal cords to try to tension the right side, holding the right cord tight. This supports his weak right vocal cord, but one consequence is that his pitch rises. He is actually speaking in a falsetto voice all the time in order to minimize the gap between his vocal folds. This is the reason for the "frail, elderly man" quality to his voice.

A scope is not a scope

When Peter's general otolaryngologist physician looked, he announced that one cord was paralyzed, but that the vocal cords were coming together. However, his otolaryngologist likely only viewed the vocal cords from one perspective – directly above and probably far away, blurry, for a very brief period of time, at a single pitch and not recorded for review as that is a typical office examination.

I move the scope closer for a detailed look, until I can align the scope more nearly parallel to the axis of the vocal cords and begin to appreciate the actual thickness of the vocal folds. In cross section, his weaker (paretic) vocal cord is very different in appearance than from above. The paretic cord resembles a linguine noodle in diameter when compared to his healthy cord, which resembles a biceps on a body builder.

Sometimes with a careful slow exam, I can insert the endoscope between the arytenoids to view the entire length of the vocal cords. At other times, I need to drip a little numbing medication on the vocal cords to anesthetize them. If I am not certain about a difference in size between the TA muscles, I have Peter sniff in. Sniffing opens the vocal cords to the fullest extent and pulls them to a long, drawn out position. In this position, his injured cord is less than half as thick as his healthy cord.

When his nerve was injured and even after some portion of it regrew, the RLN could no longer transmit a complete signal all the way to his TA muscle. The TA muscle provides the bulk in the vocal cord. With no signal immediately after the injury, there is no stimulus to the muscle. Over a period of a few weeks, Peter's TA muscle atrophied.

Yet viewed only from directly above, the vocal ligament maintains the illusion that the vocal cord is still wide. The CT muscle also pulls the vocal cord tight as compensation and helps to maintain an apparent straight edge when viewed from above during phonation. These perspective illusions hide a weak, leaky valve.

Sometimes patience rewards the observer. It is difficult for the busy physician to slow down and watch someone breathe. It can take half a minute to watch four breaths and that seems like an eternity to most examiners. I make use of that eternity and watch Peter breathe.

With only one side injured, I can watch and compare the normal side with the injured side. The good cord opens with inspiration (moving air in) and it partially closes with expiration (moving air out). We partially close the vocal folds during expiration to provide resistance to the airflow and keep the lungs more filled with air[10].

On the injured side, the vocal process (the back of the vocal cord) is fixed in position, but the middle of the vocal cord is sucked inward during inspiration (Bernoulli effect) and bows outward during expiration. This injured side is passively reacting to changes in air pressure, while the healthy side is actually tensing lightly during inspiration, so that it is not sucked in by the Bernoulli effect.

Remember, the vocal cords are a valve that control airflow for breathing as well as for making sound. Because his paretic side does not help resist the outflow of air, his lungs do not retain as much air. The alveoli (little sacs in the lung) start collapsing as they empty too completely. Then he absorbs less oxygen from his lungs. The limitation in his tennis game that he attributes to aging is at least partially due to his weak vocal cord not keeping his lungs full. The impact on Peter is that he is more out of breath with physical activity than he should be. He is also more out of breath with speaking, though that is due to the gap between his cords letting more air out during phonation.

---

10 Positive end expiratory pressure (PEEP) – think of a balloon, the more full it is, the easier it is to inflate further. If we let all of the air out of our lungs to the point that the little air sacs collapse, they are much harder to re-inflate than when they are already partially filled. We partially close the vocal folds when breathing out to maintain some back pressure in our lungs which makes filling them easier on the next breath.

In surgery I placed a plastic shim inside Peter's weak and atrophic right vocal cord – a surgery called medialization laryngoplasty or also called a thyroplasty type I. The shim accomplishes two things. It provides bulk to the paretic side so that in cross section, the paralyzed side more nearly resembles the good side in terms of mass. The paretic cord can now maintain tension without the CT muscle pulling on it.

Secondly, while an implant can move both the vocal process and the membranous vocal cord nearer to the midline, that was not necessary in Peter because his partial reinnervation had already put the vocal cord in a good position for closure. However, the implant fixed the cord in that position so that his weak LCA muscle doesn't have to hold up against the stronger left side. The weak but shimmed vocal process can no longer be budged laterally by the good cord or by an increase in air pressure. His good vocal cord doesn't fatigue as quickly with voice use now.

The outcome

Meeting Peter the day after surgery, he told me that when he phoned his daughter after the surgery, she was convinced that she was talking to her brother, not her father! Peter's voice had dropped in pitch. He no longer sounded like a "little old man."

He also noted that he could drink without being so careful. His weak vocal cord was not just impairing his conversation at the cafe, it had been impairing his pleasure of drinking (not alcohol, but any fluid). In retrospect, he realizes that he was choking on liquids more often than he realized. It had just become second nature for him to drink slowly and carefully, so it had not occurred to him to relate this to his nerve injury or to tell me about it. He also found out that he was less winded when he walked. When breathing out, his glottis could partially close now keeping more air in his lung. Lastly, he said that he could speak louder and longer.

He didn't realize that his presenting complaint of hoarseness was actually way down on his list of what he would appreciate by a restoration of his glottic valve. Being less out of breath, not choking so frequently, speaking louder, restoring a more masculine voice, a clearer vocal quality and improved volume all came with supporting his weak vocal cord.

To many people (even some surgeons), this is visually counter-intuitive. We actually narrowed Peter's airway at the glottis with this surgery and not only does the voice improve, but breathing through the smaller opening improves symptoms of being out of breath.

Some paresis is visually easy to see such as Peter's nearly immo-bile vocal cord, but we can be fooled by the patient's compensatory mechanisms that try to tense and close the vocal cords as completely as possible. The underlying reasons for the symptoms can be difficult to account for until we understand the whole problem, and a thor-ough exam can find them.

Details you may see during endoscopy that represent the "hid-den" problems:

- Squeeze of the supraglottis – the false vocal cords squeeze as a compensation for true vocal cord weakness.

- Air leak from a gap between the cords. We can best visualize this gap at low pitches which removes the CT muscle compensation.

- A difference in thickness of the vocal cords represents atrophy of the denervated TA muscle.

- A tense vocal cord will oscillate about its axis. A weak vocal cord will oscillate lateral to its axis.

- On a close exam, fasciculations of the TA muscle may be seen, supporting the diagnosis of denervation.

- During respiration, the Bernoulli effect may in-draw the weak vocal cord.

With 10 muscles in the voice box that can be injured to varying degrees, many nerve injuries are more difficult to sort out than Peter's. Not only can those muscles be injured to varying degrees, but they can reinnervate to varying degrees and there can be cross reinnervation. I am aware that the most commonly used term for this type of injury is paralysis. However, I find that the terms *paresis* and *dyskinesia* best define a neurologic injury rather than the simpler concept of no motion – paralysis. You will hear the term paralysis frequently, but since a lack of motion fails generally to describe why the voice is hoarse, why there is shortness of breath or trouble with swallowing, I prefer to define neurologic injuries of the larynx in terms of which branches of the RLN are paretic (weak) and which are dyskinetic (moving inappropriately). I will more fully explain these two terms in the coming chapter "Dyskinesia & paresis" on page 194.

Peter is now back to playing tennis and enjoying his life because of improved breathing, drinking and speaking.

Paresis – Superior laryngeal nerve

Volume and pitch impairment

Jimmy Nickel works in sales and as a singer-songwriter. He comes in with complaints about singing and loss of projection.

"I noted a sudden pain in my right neck while performing Naphtali in Joseph and his Technicolor Dream Coat. Afterwards it became painful to sing in my upper range so I avoided singing in falsetto. The worst part is that my voice is trashed anytime I have to project. If I use a cell phone, I am just exhausted. If I have to run a meeting, I'm exhausted. I now use a personal microphone with a speaker on my waist just to talk with my wife in the car and avoid projecting my voice. The muscles in my neck are always tight after I have to speak loudly."

Jimmy goes on to relate, "I first travelled to a well-known laryngologist who said I had acid reflux. I took Prilosec for several months without any benefit. Six months later I flew to another city and that laryngologist said the same thing. I took the pill again twice a day, but didn't think it would work and it didn't. Only massage has helped with the neck pain."

Listening to his voice, his upper range is very strained. For a singer, he can only reach about half of a normal falsetto range. He cannot produce soft sounds above G4, fairly low even for a male. For the next six notes, he can only reach them loudly and with great effort and C5 is the absolute limit of his upper range for this trained professional singer.

On endoscopic examination, I ask him to glide up from the lowest notes in his range. As he increases his pitch, there is some rotation of his larynx, with the back of his larynx rotating toward the right side and the epiglottis tipping over toward the right side. His vocal cords do not stretch in length as he tries to go up in pitch. The left cord oscillates about it's axis, but the right cord loses tension at his highest notes and oscillates lateral to it's axis.

We checked a CT scan of his larynx with specific attention to the cricothyroid joints and there was no visible arthritis nor visible fixation of the joints. By process of elimination, this suggests that the superior laryngeal nerve (SLN) is injured. It is plausible that an accurate electromyogram of the CT muscles would distinguish the injury.

Unfortunately, the SLN doesn't seem to grow back as fully as the RLN. Also, I know of no good treatment to compensate for a SLN paresis. The reduced vocal range seems to be a permanent injury. I cannot say whether the injury to his nerve was a physical one or whether he had some other cause such as a viral infection of the nerve.

Limited vocal cord motion

Vocal cord mobility restriction

There are three types of disorders that might loosely fit under the term paralysis if we use the very broad definition of a lack of motion:

1) After a tube is placed in someone's throat (for example during anesthesia to breathe for them), there may be ulceration between the vocal cords with subsequent scar tissue formation contracting and fixing the vocal cords close together. This may occur to a small degree even with short intubations such as during a surgical case of less than two hours duration. The scarring much more frequently occurs when a tube is left in for many days. The vocal joints may scar enough that the vocal cords cannot move with breathing or speaking. All of the nerves are still sending signals to the muscles. All of the muscles are still trying to respond appropriately. However, the scar tissue holds the vocal cord joints fixed in place.

2) After a surgery in the neck, the recurrent laryngeal nerve may be cut. No signal gets past the injury in the nerve. There is no motion of the vocal muscles with breathing or speaking. The muscles atrophy. There is no motion at all, at least in the acute condition of a complete cutting of the RLN.

3) After an injury to the RLN, the nerve re-grows, but very frequently (because the recurrent laryngeal nerve supplies muscles with opposing action) the individual fibers cross and signals to open the vocal cords go to both opening and closing muscles, so the muscles are activated, but there is no effective motion. The muscles which are stimulated simultaneously compete with each other. There may not even be atrophy of the muscles.

In all three of these scenarios often labeled paralysis, there is a lack of motion, but the etiology is different in each one, so the treatments will need to be different.

Compensation: To add complexity to this, because the function of the 10 muscles in the larynx overlap, if only one muscle is injured, the loss of motion may be partially compensated for by other muscles and on superficial exams, the vocal cords might almost appear to move normally.

Recovery: A nerve may be injured with loss of motion in a muscle, but then regrows. There is partial recovery of motion, but perhaps the muscle cannot do everything it used to do. It may fail to perform it's action completely or may only activate sufficiently under stress.

I find the word "paralysis" has too broad of a definition other than to suggest that there is some impairment of vocal cord motion. In general use, this impairment often is neurologic, but not always. If the lack of motion is from scarring, I prefer to call it fixation (and in this book I have described an example of joint fixation in the Autoimmune chapter – see "Arthritis of the laryngeal joints" on page 246 – although trauma may also fix a laryngeal joint).

Beyond that, an injury to a nerve always causes a decrease in function of the muscle it supplies. I try to identify which functions are missing, which muscles have recovered partially and which muscles are competing against each other.

Dyskinesia & paresis

Paralysis – paresis – synkinesis

Mrs. I. Kant Breeve had her thyroid gland removed 30 years ago. She developed a hoarse voice after the surgery (this happens more frequently than some physicians acknowledge[11]). Her recurrent

---

11 Why do I say "this happens much more frequently than physicians believe?" Unfortunately, and this occurs much too frequently, she was told to just wait, it will

laryngeal nerve had been injured. Stretched, cut or somehow traumatized, the nerve stopped working after the surgery and one of her vocal cords stopped moving. The cord could not move close enough to the other vocal cord to start vibrating with air flow, so she had no voice for several months.

Gradually her voice improved and eventually it recovered – or more likely it seemed to recover. She had a reasonable speaking voice for 25 years. She could no longer sing nor reach high pitches, but day-to-day, her voice was adequate for work. She accepted that as recovery.

Gradually over the past several years she started having sudden episodes of her breathing being cut off. She would be speaking and suddenly she couldn't breathe.

Every otolaryngologist who looked told her she had a paralyzed vocal cord, but that she still had "plenty of room to breathe." Several physicians told her that her episodes of difficulty breathing were from silent acid reflux and they placed her on antacids – the pills didn't help.

We should ask, do her brief, sudden, "shortness of breath" episodes (dyspnea, usually with stridor) have anything to do with her previous surgical injury? If we wanted to phrase it medically, are laryngospasms related to nerve injuries?

---

get better. Some people improve in a few days, some in a few weeks and some in a few months. Some people never return to their surgeon having been told "everything will be alright." They no longer believe their surgeon and seek advice elsewhere. This benign neglect suggests to the surgeon that most hoarseness after surgery is temporary and inconsequential.

All of the hoarseness after a surgery is from some injury during surgery and many of these injuries are nerve injuries. Some injuries are from the endotracheal tube placed in the wind pipe by the anesthesiologist. Since physicians typically don't look at the vocal folds after surgery and since the patient symptomatically recovers a voice or goes elsewhere, physicians never learn whether the patient should have been put into the "nerve injury" group or the "bruised vocal cord" group and so the surgeon comes away with a false sense of a lower complication rate than exists in reality. The surgeon tells the next patient signing up for surgery, "I rarely injure a nerve."

The word paralysis means lack of mobility. It is true that her injured vocal cord does not obviously open or close with each attempted phonation and each attempted breath. There is the implicit assumption that since her vocal cord is not moving normally there is a lack of innervation – an error. The recurrent laryngeal nerve actually has such a strong propensity to regrow that even after cutting out several centimeters of the nerve, it still often grows back to the muscles in the larynx. When the nerve grows back, the major issue is not lack of nerve input, but lack of proper nerve input.

We can say that she is suffering from a laryngeal dyskinesia. Calling her injury a laryngeal dyskinesia implies different findings and different problems than laryngeal paralysis. A dyskinesia may be present whether or not there is any observed motion impairment, though usually there is some impairment of motion. After a nerve injury the problems that result are, to some degree, due to the degree of reinnervation, but even more due to inappropriately directed reinnervation.

In the ideal world, the injured nerve would regrow back to the muscle it used to control. In the most typical severe nerve injury about half the fibers end up going to their original muscle and the other half go to the opposing muscle. Consequently, the neurologically injured vocal cord appears to be immobile. The brain tells both muscles to contract simultaneously and the net effect is that there is no motion and nothing obvious happens.

That seemed to be the case with Mrs. Breeve for many years, but something definitely changed in recent years. She began having the laryngospasms that cut off her breathing entirely for a seeming eternity (when in reality it was less than a minute, but when you can't get air, time subjectively moves slower).

On her endoscopic exam, the healthy right vocal cord opened and closed appropriately, both during breathing and during sound production. She had quite a strong voice, though I would say that it actually had a too tight, strained quality.

During the ultra-close portion of my endoscopic exam, I touched the left, non-moving or paralyzed cord lightly and it suddenly moved across the midline nearly closing off her airway. It wasn't paralyzed; it could move. It just did not move intentionally and appropriately during breathing or during phonation and it was trigger-happy. With even a small trigger, her left LCA muscle would spasm and move the left vocal process nearly all the way to the opposite cord.

This LCA spasm also increased gradually the longer Mrs. Breeve spoke. With each phonation the right cord would touch the left side and the left vocal process would move further toward the right after each touch. Then, after resting her voice and a number of breaths, the left side would relax back toward its midline, resting position.

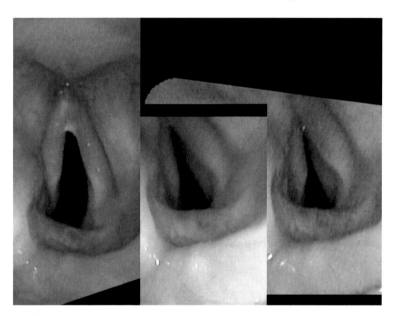

Left: *vocal cords at initial rest during inspiration.*
Middle: *after the stimulation of phonating, the left vocal process starts moving across the midline during inspiration.*
Right: *left vocal process hyperadducting during inspiration and narrowing the airway further after more stimulation.*

As both treatment and a test, I injected botulinum toxin into her dyskinetic (actually the opposite of paralyzed – hypercontracting), left vocal cord. Specifically, I put it into the TA and LCA muscles – the muscles that tense and move the cord toward the midline and closure.

Two weeks later, the opening in her larynx was larger while she was breathing. She could still make sound, though a little more softly. However, she could breathe better than she had in several years and she had not had any further laryngospasms since the injection. The paralytic effect of botulinum toxin lasted for three to four months and then the nerve regrew again and the left-sided muscles begin to hypercontract again, which she could identify because of the increasing difficulty with breathing. So Mrs. Breeve returned to the office for another treatment every few months. After several injections, she asked if there was something more permanent that could be done.

The RLN splits like a tree into different branches. It is possible to cut only the branches that go to the closing muscles (the TA and LCA) and in effect, that is what we were doing chemically with the botulinum toxin. I suggested a surgery where we would cut the anterior branch of the RLN. Then, to prevent the original RLN from growing back as it had done 30 years ago after her injury, I would route a nerve from one of her neck muscles (the omohyoid) into the cut anterior branch suppling the TA and LCA muscles. If this new nerve sprouted fibers to the muscles before the old branch of the RLN, then she would have nerve input to these muscles during phonation (the omohyoid tenses during phonation). The muscles would bulk up, and even if the muscles didn't have a completely appropriate signal to move open and closed, they would hold tension during phonation and would not tend to inadvertently spasm nor tighten during breathing in.

After cutting the anterior branch of the recurrent nerve during surgery and sewing in the donor nerve, she had a weak voice for a month and then her vocal strength began to return. The left vocal fold ultimately positioned itself near the midline leaving a much larger opening than when I first met her and she could breathe easily. She no longer had laryngospasms that would cut her breathing off and her voice was less tight and strained. She could close the right vocal cord all the way to the left and had a strong and clear voice without the tight quality she had previously.

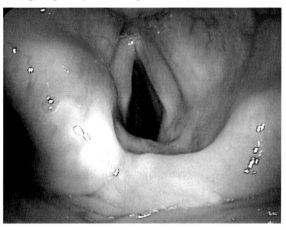

Her laryngeal opening during inspiration six months after the dener-vation-reinnervation surgery to the left side. She has a reasonably open airway with the left vocal cord now stable and set in the midline during both inspiration and phonation and very importantly, no further laryn-gospasms.

A vocal cord's muscles after a nerve injury are seldom really para-lyzed, even if there is no obvious easily recognized movement. The vocal cord doesn't lack movement though it may lack easily visible, in-tentional movement because of competing contractions. Also, it often has subtle, significant, inappropriate movement. To state it medically, after a nerve injury the vocal fold is more likely to move poorly and inappropriately (dyskinesia) than not move at all (paralysis).

For physicians who approach hoarseness after a surgical RLN injury with benign neglect, feeling that a hoarseness that recovers is inconsequential, they might wish to have their patients examined by a laryngologist to better understand their complication rate. A return to a normal voice might not mean a return to normal function.

Laryngospasm

Acute, sudden spasm

Laryngospasm is the sudden closing of the vocal cords. This is a normal adaptive response to drowning. Let's say you were suddenly submerged under water, breathed in and water began to enter the airway. As soon as the water touched the vocal cords or the lining of the trachea (windpipe) beneath the vocal cords, the vocal cords go into a spasm in a closed position. This stops the movement of air and has the benefit of preventing more water from entering the trachea and lungs. This vocal cord spasm closing off the airway typically lasts for 30 to 60 seconds before gradually opening again (though it seems like forever if you are the one not breathing). Presumably that would be enough time for you to reach the surface. As the vocal cords open ever so slightly, breathing in is very noisy as air passes through the tiny slit (the Bernoulli effect).

Since we don't presumably spend a lot of time drowning, most of us find the inadvertent laryngospasm – perhaps from swallowing while talking – merely a startling inconvenience.

However, there are people like Mrs. Breeve who suffer from frequent, recurrent laryngospasm, some as often as several times per day. My experience is that laryngospasm more than once or twice a year is likely secondary to dyskinetic reinnervation after a nerve injury. While Mrs. Breeve had an initial severe injury, even a minor nerve injury seems to sensitize the larynx to more frequent and inappropriate laryngospasms.

Whenever I see a patient with frequent laryngospasms, I look for evidence of a nerve injury. If I find it, I consider the use of botulinum toxin even into a vocal cord that is not grossly moving and may have been previously considered paralyzed, since in all likelihood the issue is really dyskinesia or hyperkinetic movement with any stimulation of the injured side.

Sometimes a single botulinum toxin injection will relieve the recurrent laryngospasms. Sometimes the botulinum toxin injections need to be repeated. If repeat botulinum toxin injections are helpful, one can consider a reinnervation procedure with a different nerve supply to block the spasms permanently.

Bilateral vocal cord paresis

A rock and a hard place

Mrs. Abigail Gris awoke from her neck surgery with a weak voice. Dr. Manjoo, her surgeon, says, "Try not to worry, this happens sometimes. The breathing tube in your throat may have caused a little swelling. Just wait, it will get better."

Sound reasonable? Perhaps plausible, but is it true?

When Mrs. A. Gris didn't improve after three weeks, she called Dr. Manjoo back. He said, "You better go see an otolaryngologist to have a look at your vocal cords. I have never nicked the nerve to the vocal cords, but it is possible that I stretched the nerve a bit."

She went to an otolaryngologist who looked in her throat with a scope. He said she had a paralyzed vocal cord, but it looked like it would get better. She should just be patient.

Over the next several weeks, her voice started to recover and in several months it seemed to be back to normal. She and her physicians assumed this was the end of her story.

Do surgeons nick the nerve to the vocal cord? I doubt they make an actual nick or cut the nerve very often. But certain surgeries put the main nerve, the RLN, at risk for injury. I suspect the injury is most often a stretch injury, but the effect of a stretch is initially the same as a cut nerve. Otolaryngologists; orthopedic and neurosurgeons who operate in the neck; endocrine surgeons who operate on the thyroid and parathyroid glands; chest and thoracic surgeons who operate on the lungs; heart surgeons who get near the junction of the aorta with the heart and vascular surgeons who operate on the carotid artery all put the nerve to the vocal cords at risk when they operate near the RLN.

It is also true that an anesthesiologist putting a tube in your throat can also injure the RLN. There is a cuff on the tube placed in your trachea that the anesthesiologist inflates to seal the airway. If the cuff is just barely below the vocal cords and if the cuff is inflated too tightly, the anterior branch of the RLN can be pinched between the cuff and the cartilage of the larynx. The injury will be to the adductor muscles (TA and LCA) and the vocal cords will not close well afterwards.

The anesthesiologist can prevent this problem by ensuring the cuff is far beyond the vocal cords, but more importantly, if the anesthesiologist only puts just enough air in the cuff to stop an air leak, and no more, the nerve should not be injured from excessive pressure.

Mrs. A. Gris did just fine for five years. Inauspiciously, she again developed pain in her neck. Having moved to a new city and state, she found a new neurosurgeon who felt he could help her pain by operating on her neck again. He chose to approach the neck from the opposite side (I do not know the reason for this decision). Perhaps he heard her good voice and assumed that her vocal cords were working normally and he felt that he could avoid the scar tissue from the previous surgery.

She awoke from this second neck surgery with no voice again. Over a few weeks her voice started to return, but it was still weak when I first saw her. The right vocal cord, previously injured was seen to be in a nearly fixed position near the midline. The left vocal cord, newly injured, could move slightly apart with breathing in. It could move slightly toward the midline with attempted sound production. If she breathed in quickly, the Bernoulli effect would pull in the weak left cord, generate noise (stridor) and make breathing more difficult.

Although her primary concern at this visit was to get a stronger voice back, I asked her to wait and see what spontaneous regrowth would bring. Unfortunately, the regrowth of the injured RLN, while very helpful in a one-sided injury, usually becomes a significant problem when the same regrowth happens on the second side. Over the next several months, her voice did, indeed become stronger as the nerve grew back. Breathing also became more difficult. She saw me again four months after her first visit.

On endoscopy, when she would actively breathe in, the recently injured left cord would actively close. Enough of the fibers that normally opened her vocal cords during inspiration (via the PCA muscle) had grown into the muscle that closes the vocal cords for speaking (LCA muscle) such that, the deeper she breathed in, the more strongly the left cord closed.

She now has a dyskinetic reinnervation. The vocal cords are not moving a lot, so superficially they might appear paralyzed, but on our close examination, this small, but very significant movement closing her left vocal cord is actively shutting down her breathing. Six months after her second RLN injury she has to avoid or minimize any activity or she cannot breathe well. The more she tries to do, the less air she gets. Her voice has become loud enough now, although it has a tight, lightly spasming quality. She lacks fine control over her voice.

Her options have become very limited. The simplest surgical solution is to place a tracheostomy tube. A tracheostomy is a hole in the neck with a plastic or metal tube inserted to keep the hole open. This allows breathing in and out through the neck bypassing the vocal cords. A one-way valve can be placed on the tracheostomy and air can enter the trachea on inspiration through the tube, but is then directed out through the vocal cords for generating sound or voice on expiration. Then she can both breathe and speak.

There are other options, such as permanently cutting or partially removing one of the vocal cords or tying one of the vocal cords out into a permanent open position. One can partially or completely remove the joint (arytenoidectomy) that opens and closes one of the vocal cords to enlarge the airway.

However, there are several problems with creating a fixed opening into the windpipe. The voice becomes softer the larger the opening. Although she will get air in, more of the air will leak out as well, and she will likely be more out of breath with speaking and activity. It is not a restricted breathing, but rather it is an "out of breath" sensation. Lastly, since the vocal cords are a valve, not only for breathing and voicing, but for swallowing, the larger the opening into the windpipe, the more that liquids and food go down the wrong way. So coughing while eating increases and pneumonia becomes more common, since food in your lungs is not very healthy.

Healthy vocal cords are an active valve that dynamically open and close to regulate airflow and food flow. With a fixed valve or a malfunctioning valve, such as with her bilateral RLN injury, one has to make choices between a completely closed valve (no air movement, so not very functional), a completely open valve (no sound production and lots of food in the lungs, so not very functional), or an in-between, leaky valve. There are no good choices, just bad ones with a lot of compromises.

Mrs. A. Gris started by getting a tracheostomy. She really didn't like this option, but it very quickly allowed her to breathe very easily. In order to see if we could get the tracheostomy tube out, we tried to cut the anterior branch of her recently injured nerve and grafted a new nerve onto that branch. It opened her airway slightly, while still preserving her voice, and it allowed her to function with a one-way valve on the tracheostomy tube more easily.

She still wanted to try and get the plastic tube out of her neck, so we started to remove portions of her left vocal cord and left arytenoid cartilage. She could now plug her tracheostomy tube for extended periods of time and her voice was a little softer, but there were times with exertion where she had to open the tracheostomy tube to get enough air. Consequently she could not do without the tracheostomy tube completely.

She started to get recurrent pneumonias from small pieces of food and liquid that would get through her now partially open vocal cords. We had reached a limit to what I could do. If I opened her vocal cords further, she could have the tracheostomy tube removed, but her voice would become more of a whisper and the pneumonias would likely increase. She appears to be stuck with a tracheostomy tube in order to breathe.

If you have ever have a RLN injury, before considering any additional surgery on your neck or chest, you should have a good laryngeal exam to assess the risk of an additional surgery. Even with a normal voice, it is possible to have one vocal cord that does not move well or at all. If that happens to be the case, then before another surgery near either RLN, an updated exam will allow you to understand the incredibly high risk to operating anywhere near the remaining uninjured nerve. Operating near the already injured nerve is a small risk. Operating near and injuring the nerve on the opposite side ends up as a significant, life-altering event.

I consider this analogous to our vision. We have two eyes which offers some redundancy.Losing one eye means a loss of depth perception, but we can still see to read, drive and look at the smiles on our friends' faces. But losing the second eye leads to complete blindness. We are surely still alive, but we completely alter the way we live. The RLN supplying the vocal cords is the same type of partially redundant system, very forgiving when one side is injured, unforgiving if the second side is injured. Mrs. A. Gris could tell you how life altering this injury is.

Laryngeal dystonia

Adductor spasmodic dystonia

At age 83, Effie Forte reported one year of effortful speaking. If she shouted for awhile, she could continue to speak with effort. After resting her voice though, it was very difficult to get her voice started again. She spoke in short, tight bursts when describing her problem.

On endoscopy, I noted that she squeezed her larynx tightly from each side every time she tried to produce a sound and that sound would be cut short by the squeeze. Except for the severe squeeze, her vocal cords appeared normal and using stroboscopic light, they vibrated normally for the brief period of each vocal effort.

I felt that her diagnosis was adductor spasmodic dystonia (also commonly called adductor spasmodic *dysphonia*). Adductor spasms of the vocal cords mean that the vocal cords over contract, irregularly closing tightly when trying to phonate. It is a focal dystonia, which are spasms of a single muscle or a group of muscles. Spasms tend to be irregular and occur during intentional use. They usually do not occur at all when the muscle is at rest.

While a tremor is also an over-contraction, often occurring only during intentional movement, a tremor has a regularity about it. As one muscle over-contracts, the opposing muscle takes over and an oscillation starts as each opposing muscle contracts in turn. These contractions tend to be regular and occur typically about 4 times per second. They are perhaps most well visualized when a person with a hand tremor tries to pick up a coffee cup. If the cup is full, the coffee slushes out each side of the cup. We call this an intention tremor since when the person is not trying to do something, the tremor may become unapparent. It occurs with intentional movement.

Spasms of dystonia also occur during an intentional movement, but there is no counteracting muscle contraction. Spasms can be a nearly continuous contraction or a very intermittent and irregular contraction.

In the larynx, spasm of the closing muscles during phonation, *ADductor* spasmodic dysphonia (AD SD) is the most common type of laryngeal spasm and causes the voice to cut off during speech as the vocal cords clamp shut. *ABductor* spasmodic dysphonia (AB SD) is less common and when trying to phonate, the muscles inadvertently open, causing breathy breaks in the speaking voice.

AD SD is one of the few laryngeal disorders that is often more easily visualized during speech than during a steady sound. A person with AD SD can overpower the closure of the cords and emit sound, but this is very fatiguing. Mrs. Forte had such a tight squeeze that it was like a single strong spasm every time she spoke.

I recommended botulinum toxin injections to weaken the muscles closing her vocal cords. No one knows what causes a dystonia, though I suspect the problem is in the brain, not in the nerve endings. However, for now, the best available treatment seems to be partially blocking the nerve input to the spasming muscles. It seems reasonably safe to do so with botulinum toxin which wears off if the result is not beneficial. The wearing off is also a downside as the beneficial effect wears off and the injection needs to be repeated in three to six months.

Mrs. Forte was not very excited about needing repeated treatments several times per year. The injection is fairly quick, but unlike a pill for a disease that can be taken at home, most people find it difficult to place a needle into their own neck, find the vocal cord muscle and insert the medication. Thus, the treatment requires a trip to the doctor, a higher cost because you have to pay the doctor as well as pay for the medication and, while brief, there is the not terribly pleasant idea of having a needle inserted into your neck.

She wanted a cure, not a treatment, but agreed to try the medication. She was able to speak more freely after the treatments for three to nine months depending on the dose. However, after three treatments she still sought a cure, not a treatment. I felt we learned something by the treatments. She was improved after each treatment, suggesting at a minimum that the diagnosis of AD SD was correct.

I didn't hear from her for awhile, then about 18 months later she returned. Speaking was very effortful again. Her voice was worse than before. On endoscopy, now her vocal cords had lost some of their white color. Gradually I learned that in the interim she had gone to a general otolaryngologist who told her that she had Reinke's edema. He had not taken any photos, but he had taken her to the operating room and had "stripped" her vocal cords to remove the edema. Her voice was worse after the surgery and had not improved with time as the surgeon had predicted.

"Reinke's edema" is a condition also known as "smoker's polyps" (see "Reinke's edema" on page 112), which tobacco smokers who talk a lot develop. In fact, only tobacco smokers who talk a lot develop them. There is something about the combination of smoke and the vibration of the vocal cords that causes the edema. Besides that, the only complaint patients with smoker's polyps typically have is "Do I have a cancer?" The change in their voice is a deeper pitch because the vocal cords are thicker. Thick cords vibrate at a lower pitch just like the thicker strings on a guitar or piano. However, the females and the males I have met with smoker's polyps not only aren't concerned by their pitch, they often like their deep pitch, even the women who are so deep they are mistaken for men on the phone. They just say that everyone who knows them recognizes their voice. If they don't have cancer, they don't want to get rid of their smoker's polyps. Imagine a female jazz club singer and you likely have the sound of smoker's polyps in your mind.

Mrs. Forte did not talk a lot, did not ever smoke tobacco, did not have a deep voice and I had photos of her vocal cords from one-and-a-half years ago that were normal in terms of their structure.

One of the problems for physicians is that their word carries as much weight as a photo or video recording. We quite often answer only to ourselves. Since the physician who diagnoses the patient also treats the patient, the diagnosis and the treatment always seem to line up in a report. I really don't know how to explain this particular physician's diagnosis of Reinke's edema in a women who did not ever smoke and who on my video recordings had normal vocal cords only 18 months ago – error, zealousness, wanting to please the patient? Now the patient has two problems, stiff vocal cords (from the stripping) and spasmodic dysphonia. She is now in a tough position.

Botulinum toxin injections may not be the only treatment for some laryngeal dystonias. When the spasms involve the adductor branches of the recurrent laryngeal nerve, then surgically cutting the adductor branch of the RLN and hooking up a sacrificial nerve from another neck muscle (ansa cervicalis) is also an option. This procedure is called a DEnervation-REinnervation surgery (DeRe) or a selective laryngeal adductor denervation (SLAD) (physicians love complicated, if accurate, names and their associated abbreviations). However (there is often a however), her insurance will not cover this procedure, labeling it "experimental." It is also expensive.

She has public health insurance designed for the poor. Interestingly, the insurance paid for the vocal cord stripping (remember the diagnosis and the treatment lined up on a list at the insurer so approval was given). It would cover the botulinum toxin injections with enough paperwork on my part, but it wouldn't cover the DeRe surgery.

She continues to live, speaking with great effort and now with a poor quality voice when she is able to get sounds and words out. Many times I ask myself, what is the value to a patient and to society of a correct diagnosis?

Abductor spasmodic dystonia

In his mid-fifties, Bob Icon has had to reduce his preaching schedule to a minimum. For the past 15 years his voice has become progressively more undependable. His voice fades out, breaks up and sounds unsteady. It requires a great deal of effort to speak. The first ENT he saw diagnosed vocal polyps and sent him to speech therapy. However, the therapist told him that his voice sounded like he had abductor spasms. He went to a second ENT who tried injecting botulinum toxin using a ultrasound for guidance, but each time it made his voice breathy for awhile without improving his voice. At one point, he had a cold and his voice improved temporarily during the infection. He was sent to a neurologist who gave him clonazepam. This made him sleepy and made his snoring worse and didn't seem to help his voice.

Abductor laryngeal dystonias are spasms of the opening muscles of the larynx, the PCA muscle. The patient has "breathy breaks" in the voice while speaking. This type of dystonia may be treated with botulinum toxin injections into the PCA muscles.

Although he states that one physician tried to inject the PCA muscles, it sounds like she probably never actually placed the medication in the correct muscle based on Bob's description of the injection. He described what sounded like a TA muscle injection and he had breathiness after the injection – a typical side effect of a TA injection. Since the PCA muscles open the vocal cords for breathing, they will not open the vocal cords as far if they are weakened by botulinum toxin. Vocal cords that cannot open well can impair breathing, and with the vocal cords closer together the voice should be stronger. If too much botulinum toxin is placed in the PCA muscles, the patient will make a noise with every breath inward because they cannot open the vocal cords far enough apart to prevent vibrations from the Bernoulli effect. If too little botulinum toxin is placed there will be no effect. He had neither of these effects.

The fact that he had a weak, breathy voice after the injections suggests the botulinum toxin was actually in one of the vocal cord muscles that tighten the vocal cords, such as the TA or the LCA muscle. Whether that was intentional or inadvertent on the part of the physician, I couldn't know with the information I had.

I discussed with the patient that different physicians use different techniques for injecting the PCA muscles. Some physicians try to pull the larynx forward, which is uncomfortable, and pass a long needle behind the larynx. I utilize an EMG and I pass the needle straight through the front and out the back of the larynx so that I know when I am encountering the PCA muscle by the feel of the cartilage and the EMG signal. I also told him not to expect breathiness as a side effect, rather there should be some mild restriction in his breathing, possibly noisy breathing.

Part of my diagnostic process is to place the botulinum toxin and see the patient back in two weeks and see if the intended muscle is weakened. If the muscle is not weakened, then we would try a higher dose. If the patient is not improved, despite weakening the intended muscle, then we would need to rethink the diagnosis. He was, however, improved and has resumed his preaching as well as continued with the injections.

Respiratory dystonia – inspiratory

For several years Mr. S. Tate Seller has noted pain in his shoulders and neck, and for two years he has noted difficulty talking and breathing. When he tries laying down to sleep, he struggles for about two hours before falling asleep because of difficulty breathing in. His breathing seems better in the morning, but as soon as he starts talking or moving, the difficult, noisy breathing returns. His voice is normal, but his breathing is very noticeable, gasping between words, especially on the phone. He is an estate broker, and when people hear him gasp he loses their confidence as well as the sale. He cannot even sell enough to pay for health insurance anymore. When stressed, his breathing is more restricted and he has to continuously clear his throat.

Chiropractic treatments and massage help reduce the neck and shoulder pain temporarily, but have not improved the noisy breathing. His family doctor diagnosed him with anxiety (try not being anxious when every breath is noisy and difficult). He lost weight, but the anti-anxiety pills did not improve his breathing. He was told it was due to allergies. Allergy medication did not help, but clonazepam, a muscle relaxant, did improve him a bit.

On endoscopy, I note that he can speak easily and that his larynx is smooth and stable during phonation. However, when he attempts to breathe in, instead of opening, both vocal cords appear very unsteady as he tries to breathe in, spasm closed and create quite a bit of noise during inspiration. The vocal cords never really open anywhere near normal except briefly during a cough.

While the more common dystonia of the adductor muscles occurs during attempts to speak, his spasms occur during attempts to breathe. Dystonias typically occur during a task or intended muscular activity. They are an irregular activation where the muscles contract much more than they are supposed to. Other perhaps more familiar examples include torticollis, where a large neck muscle contracts pulling the head sideways; blepharospasm, which causes the eyelids to nearly continuously blink and writer's cramp, which occurs in the hand as fingers cramp up while trying to write or play an instrument. Mr. Sellers's LCA muscle is spasming when he tries to breathe in, an exhausting condition. By the time I see him, he has to sit up in bed struggling for every breath until pure fatigue renders him asleep.

I inject botulinum toxin into the closing muscles of his vocal cords and one month later when I see him, he has stopped all of his nasal inhalers and all of his allergy medications because they were unneeded. Speaking is easier, breathing is easier, though he still audibly gasps between words. Sleep is also coming on easier.

On future injections, we place larger doses of botulinum toxin and he breathes better, falling asleep easier. After a number of treatments, between injections he never returned to as severe a condition as when I first saw him, even if he goes for a long period between injections. There seems to be some partial permanent improvement after several injections. Mr. Seller receives injections about every six months.

Respiratory dystonia – expiratory

Mrs. Tock complains that she has too much phlegm, a rather common complaint in an ENT office. During endoscopy, there is no excess of secretions but there is an unusual movement of her vocal cords during breathing. Her vocal cords open to take in air, but then close prematurely and sit against each other shortly after she breathes in. Normally, vocally cords close slightly during expiration (breathing out) to provide some resistance and keep the lungs from collapsing. Hers are closing most of the way together, such that the flexible part of the vocal cords are completely together. I find that many times the sensation of excessive mucous production relates to only a small amount of mucous which ends up sticking to the vocal cords. For Mrs. Tock, with the membranous vocal cords sitting against each other she senses the touching as phlegm build up, but it is actually each vocal cord touching the other that she is sensing. There is no extra mucous. She has spasms of her LCA muscles occurring during expiration, so this is a respiratory dystonias of expiration.

An injection of botulinum toxin into the adductor muscles softens her voice, but also allows the vocal cords to rest apart during breathing and relieves her sensation of phlegm build-up for several months.

Other dystonias

Muscles surrounding the larynx, muscles supporting the larynx and muscles involving the speech tract above the speech line (pharynx, tongue, lips, jaw) may spasm and impair airflow and thus impair voice and speech. These dystonias may interrupt or alter speech and so may sound similar to some of the above laryngeal dystonias that involve specific vocal cord muscles. Some regional dystonias respond well to botulinum toxin injections, particularly the supraglottis.

With muscles more closely involved in swallowing, side effects from botulinum toxin weakening swallowing can make treatment of the muscle spasms not worthwhile for the patient.

Summary of dystonias

Dystonias can involve a single muscle or a local group of muscles. They occur as a spasm during attempted voluntary movement, effectively an inappropriately strong movement when not desired. Whether the spasm occurs during breathing or speaking, they also may involve some of the nearby muscles in the neck which are recruited to help move air through the tightening vocal cords. This caused Mr. Seller to have neck pain from chronic muscle tightness. This neck pain also improved after each laryngeal treatment.

Dystonias of the larynx most frequently occur with phonation, typically the TA or LCA muscles, interrupting and cutting off the voice and consequently interrupting speech. Spasms of the PCA muscle during phonation cause inappropriate opening, which in turn causes breathy breaks in sound production. Depending on the timing of spasms of the TA or the LCA muscles or both, interruptions can also occur with breathing in, cutting off the airway as with Mr. Seller's or with breathing out as with Mrs. Tock.

Vocal tremor

Highly regular inappropriate movement

At 87, Elizabeth Longlife comes in with her children. The children complain that Elizabeth is getting more difficult to understand over the past five years. She sounds unsteady to them. Ms. Longlife states that she is not bothered by her voice but came in at the request of her children.

Unlike the irregular spasm of AD SD, a vocal tremor is a regular oscillation of the larynx giving the voice a very shaky or unsteady sound. Frequently, a vocal tremor does not bother the person with it. Mostly it bothers those who have to listen to and understand the person with the vocal tremor.

Generally, a patient with spasms senses a great deal of effort and feels they are fighting against the spasms. A patient with tremor does not sense any significant increase in effort with making sound. Her voice is just unsteady.

Vocal tremor can be mitigated to some degree with medications that work for other types of tremor. The most common medication is propranolol, which will tend to decrease the intensity of the oscillations but not eliminate them. Mrs. Longlife was happy to know that she didn't need any more medications unless the tremor was to the point that it bothered her.

Neurologic combinations

Spasms and tremor are associated

It is not uncommon to have more than one of the above types of tremor or spasms. About 10 percent of people with AD SD also have a vocal tremor. A person may have both adductor and abductor spasmodic dysphonia. A patient may have a vocal cord dystonia and a neck dystonia. In a person with a mixed neurologic voice disorder, I typically begin treatment for one of the disorders at a time so that both the patient and I may decide how effective each treatment is alone before combining them.

Tumor

Benign laryngeal tumors

Leukoplakia

Jane Smith is in her mid-fifties. She has smoked about 10 cigarettes a day for the past 10 years. She developed a hoarse voice four months ago, though perhaps she lost some of her higher singing notes before that. She was treated with Nexium for awhile and when she did not improve, she had a biopsy. The irregular area on her left vocal cord proved to be squamous cell carcinoma. When a white spot reappeared on her vocal cord shortly after the surgery removing this, she came to visit me.

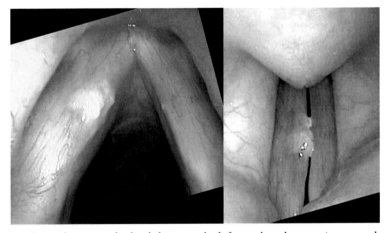

Left: *A white spot - leukoplakia - on the left vocal cord appearing several weeks after removal of a small vocal cord cancer (a squamous cell carcinoma).*
Right: *Because the white patch prevents complete closure of the vocal cord, air leaks around the white patch causing a husky hoarseness. When she sings the two segments vibrate at two different pitches giving her a rough hoarseness (diplophonia) also.*

We went back to surgery and re-excised the area of leukoplakia again. This time the biopsy was benign. There was no cancer, only thickening of the mucosa. Her hoarseness went away since the white spots no longer stood out from the edge of the vocal cord as much. A smaller pair of white spots came back after surgery. Rather than cut them off again, we watched the white spots with regular examinations over the next three years.

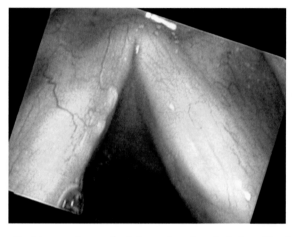

Two small white patches on the left vocal cord with feeding blood vessels.

Although relatively stable in size over several years, when Ms. Smith's spots of leukoplakia seemed to be slightly enlarging again, we decided to treat them, this time in the office with a new technology, a flexible pulsed KTP laser. Treatment with the KTP laser effectively destroyed the blood supply to this area and burned off the white spots so there was no specimen for a pathologist to look at. I really cannot say this time whether the leukoplakia was benign or cancerous except that it grew rather slowly for three years. After treatment of both the leukoplakia and the feeding blood vessels, the white spots have not returned, although we continue with regular examinations of her vocal cords. We could treat the leukoplakia again if it comes back.

Leukoplakia may be benign or at times it may represent a cancer. After identifying abnormal white spots on the vocal cord, options for patients include treatment that removes the white patches or makes them go away or careful observation and removing any white patches that grow over time.

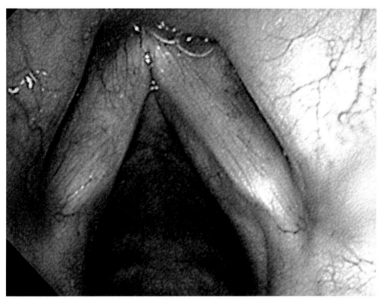

Six months after treatment with the KTP laser, the leukoplakia has not returned. This photo is taken with a high definition camera using a color-altering technique to emphasize the blood vessels.

Careful close observation, with carefully directed precise treatment when necessary, may be a successful substitute for what, in the past, was essentially overtreatment. Cutting away part of the normal vocal cord along with leukoplakia, because it might be a cancer, might have the same cure rate, but sacrifices voice quality when a significant amount of normal tissue is removed.

Malignant laryngeal tumors

Smoking and squamous cell carcinoma

Stefan Kowal developed hoarseness two years ago. He had been seen twice by a laryngologist who told him he had a polyp on the right vocal cord and treated him with antibiotics, reassuring him it wasn't cancer. His vocal quality fluctuated, such that at times he thought he was improving. Then he definitely became more hoarse. He now cannot yell or even be heard well with any background noise and he develops pain if he speaks very much.

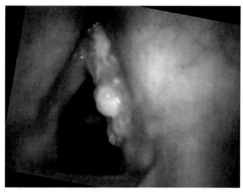

A rough, irregular textured surface on the right vocal cord with white and red areas is rather suggestive of cancer.

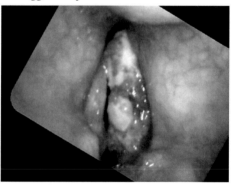

During closure of the vocal cords, white patches (leukoplakia), a pebbled surface and growth above the level of the opposite vocal cord are suggestive of cancer.

Physicians may over or underestimate the possibility of cancer. Some, who do not want their patients to be concerned, might underestimate the risk of cancer, others bring up the probability of cancer such that surgery seems like the smaller risk than the uncertainty of not knowing. If I bring up the possibility of cancer with each patient for almost every bump, I could remove something from almost every one of these patients.

Sometimes a discussion including the potential of cancer leads to patients no longer processing additional comments during an appointment. All the patient can think about is, "I have a cancer."

In Mr. Kowal, I was in awe how such a large mass, covering the entire surface of one vocal cord, could be examined two years in a row and still be treated as a polyp. I suspect that equipment played a role, as the flexible fiberoptic scope found in many offices presents a blurry picture. There was no recording made of his prior exams; quite possibly the image in the eyes of the otolaryngologist was fleeting. Couple the problems of no recording and a blurry view with the possibility that the physician did not move the endoscope very close to the vocal cords and the delay in diagnosis is now understandable, even if not desirable.

Despite its rather extensive appearance, this cancer proved to be entirely on the surface and was peeled off of the right vocal cord with a CO_2 laser. One month after the surgery he still had a slightly elevated red area on the edge of the vocal cord. While this appeared to be a granuloma on my endoscopic examination, Mr. Kowal now had an elevated concern for any finding that might represent cancer. Since lesions with blood in them respond well to the office pulsed KTP laser, we treated this and the red lumps were gone by his next office visit.

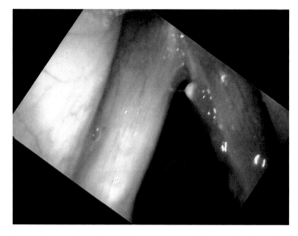

A granuloma formed on the right vocal cord during healing. We used an office KTP laser to shrink it all the way down.

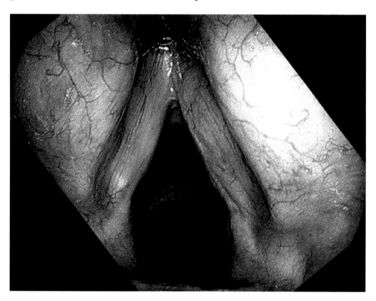

Six months after removal, the right vocal cord is smooth and straight.

Even though Mr. Kowal's right vocal cord is now scarred and does not vibrate very well, the right vocal cord has a smooth surface that the left vocal cord can close against and his voice seems to him to be as strong as ever. He returns to the clinic for frequent check-ups and the tumor has not recurred as of writing.

Follow-up examinations are important in any patient with cancer. For me, follow-up examinations are equally as important as the diagnosis. The examinations are essentially part of the treatment.

Because of the fear that the word cancer strikes in patients and physicians, there is a tendency to overtreat. If this over treatment is benign, there is no problem. However, almost no treatment is completely benign. For example, the more tissue a surgeon removes, the more likely he is to remove the cancer completely. However, there is a diminishing return to this approach. Removing Mr. Kowal's entire larynx with a total laryngectomy would have been equally effective treatment. More certainly, it would have removed all of the cancer. However he would now be functioning without a voice box.

When I removed Mr. Kowal's tumor, I cut right on the interface of the tumor and his normal tissue. When the pathologist looked at the cancer that I removed, he said that the cancer extended all the way to the edge of my excision. In the past, standard treatment has been to remove some good tissue as well, just for reassurance. The pathologist then typically says that all the margins he looked at are negative for cancer.

For me, the reason that follow-up examinations are as important as the diagnosis is that I have cut closer to the cancer than many other surgeons. This leaves more normal vocal cord in place. With regular, close examinations and with high-quality endoscopes, I believe that I will see any remnants or recurrences of the cancer early and can treat them appropriately while still very small.

This avoids the problem of surgical over-treatment that otherwise removes some normal vocal cord and makes the patient's voice worse. This also avoids the problem of over-treatment with radiation therapy and chemotherapy. While radiation therapy is just as successful at treating small vocal cord cancers, normal tissue is exposed to the radiation and complications, such as a dry throat, are unnecessary and lifelong.

Distant tumors

Paralysis as an indicator of problems elsewhere

In her mid-seventies, Mrs. Adell Jones noticed her husband becoming more hard of hearing over two to three years. She ponders out loud to me, "Perhaps I am the one having trouble and perhaps postnasal drip is ruining my voice." She also concedes, "Not only is my voice weak, but I am having some difficulty speaking. I have given up talking to my friends on the phone. My voice is too unsteady for them to understand me. The more I talk, the more my voice fades."

While she is telling me her story, I notice that her voice is cracking and jumping up to higher pitches. Some of her words are not articulated clearly. She can only make a sound at her regular speaking pitch for about eight seconds on one breath. She has problems both above and below her speech line.

I begin my examination in her mouth and I notice that the left half of her tongue has begun to shrink in the size. She also has poor control of her tongue and the left side of her tongue is weaker when she sticks her tongue out. I can see that she has indications of a neurologic problem that involves not only her voice but also her speech.

I place the endoscope into her nose and part of my neurologic examination of the larynx always includes the palate, since the Xth cranial nerve[12] supplies the muscles along the side of the throat and the palate as well as the larynx. The left side of her palate is weak and does not elevate when she makes sounds such as /p/, /t/, or /k/[13]. Air escapes out through her nose on these sounds, although it is difficult to hear since her voice is so weak. As I move the endoscope into the

---

12 Traditionally the cranial nerves are numbered as Roman numerals so the 10th cranial nerve might be abbreviated as "CN X".
13 Slashes around a letter or group of letters is a visual way of representing a sound.

throat (pharynx), I ask her to make a high-pitched sound and the left side of her throat does not contract. Finally, when I look at the vocal cords, the left vocal cord is not moving and the muscle within the vocal cord has atrophied.

Whenever one vocal cord is paralyzed and there is no obvious reason for the paralysis, one of the considerations is that a tumor may be putting pressure on the recurrent laryngeal nerve. Additionally, whenever there is a neurologic problem in the larynx, the examiner needs to assess the other cranial nerves in the vicinity. As the cranial nerves wind their way from the brain to the muscles that they control, they are sometimes close to each other and in some locations far apart. If more than one cranial nerve has a problem, the astute examiner will look in the location where the cranial nerves are near to each other.

In Mrs. Jones, I find weakness of the left side of her throat and palate suggesting that the Xth cranial nerve, which supplies the pharynx, palate and vocal cords, is not working at all on her left side. When I add in the finding of the left-sided tongue weakness, indicating at least a partial paralysis of the XIIth cranial nerve, I know that both of these cranial nerves (X and XII) pass right next to each other as they exit the skull. This clue makes the base of the skull the most likely location to find Mrs. Jones's problem.

Before dealing with her vocal cord weakness, I obtain an MRI scan of her skull base. I find a tumor compressing the Xth and XIIth cranial nerves in this location.

A nerve paralysis can be an indication of an injury and consequently of a tumor anywhere along the path of the nerve between the brain and the vocal cords. On the left side, the Xth cranial nerve follows a rather convoluted path. After it leaves the skull it passes all the way down to the heart and around the aorta, the main blood vessel leaving the heart. Consequently, the Xth cranial nerve can become weak if pressure is put on it, such as from a dilation of the vessels of the heart, from tumors in the lung or from tumors of the thyroid gland.

Unusual laryngeal tumors

Any cell can become cancerous - cartilage

John Smith notes his voice has been worsening for the past three months, and is now losing it completely at times. Recently he has more difficulty breathing as well.

"I feel like I'm breathing through a straw," he complains.

He smoked cigarettes for 30 years and he is finishing treatment for a tumor of his spine that seems to have completely gone away according to his oncologist's most recent evaluation.

Listening to his voice, it is deep pitched. He has almost no vocal range. He has almost no volume and runs out of breath after six seconds of making a sound[14].

On endoscopic exam, the tissue beneath his vocal cords is squeezing in from the sides narrowing his airway. This is the area called the subglottis.

---

14 Most people feel a shortness of breath if the maximum phonation time is less than 10 seconds. While normal varies a great deal, a healthy young person should be able to generate sound for 20 to 40 seconds.

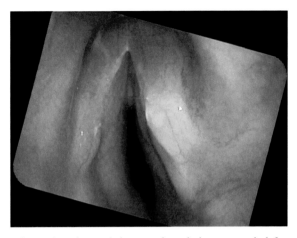

The cricoid cartilage beneath his vocal cords has expanded from tumor growth, narrowing his airway to a slit beneath his vocal cords even as he tries to fully open his vocal cords.

In this case, further testing with a CT scan shows that the foundation of his larynx, the cricoid cartilage, is almost completely replaced by tumor. A biopsy reveals a giant cell tumor of the cartilage.

Hemangioma

Hope Rouge, currently 50, recently had a surgery for snoring and was told by her ENT at that visit that her voice box looked abnormal. She came to see me. Although not a big concern to her, she says, "For about one year I have noticed some hoarseness. I sound like I have a cold all the time and my voice cuts out if I try to yell."

She smoked a pack of cigarettes a day for 25 years but she quit smoking seven years ago. Her voice is deep sounding, almost masculine in quality. She says that she sang quite well when she was young.

When I test her voice, sound starts cutting out from about middle C upward. It sounds like something is touching her vocal cords. On endoscopic exam the edges of her vocal cords are normal. However, she has deep red, dilated blood vessels filling and enlarg-

227

ing the right false vocal cord. There are also a few on the left side. When she tries to make a sound, this hemangioma[15] presses against the true vocal cords and stops the vibrations. We order an MRI scan and there is a vascular lesion, that is, blood vessels are filling the right false vocal cord.

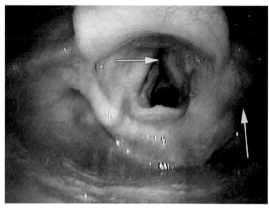

There is a bulge of blood vessels filling the right false vocal cord (arrows) both inside and outside of the aryepiglottic fold.

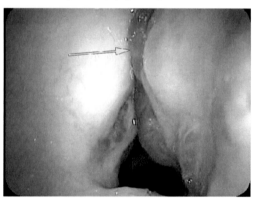

Close-up view of blood vessels completely filling the right false cord (arrow). There are a few small vessels on the left side as well. The mass of vessels is so large that we cannot even see the right vocal cord nor the front half of the left vocal cord.

---

15 Hemangioma: A benign tumor consisting of the cells making up the walls of blood vessels.

Now, five years after a surgical excision, she still had small remaining blood vessels on the right side but no further symptoms. She will continue to be examined with an endoscope on a regular basis or if any laryngeal symptoms return. Additionally, since her treatment, a new medical approach has been discovered where some blood vessel tumors will shrink with medical treatment with a beta blocker.

Although the mucosa, which is similar to skin (epithelial cells), is the most common cell type to become cancerous, any cell type in the larynx can become malignant. This includes nerve cells, blood vessel wall cells, fibrous cells, muscle cells or in the previous two example cases, cartilage cells or blood vessels.

Mucous gland disorders

Plugged mucous glands

Mucous glands in the cords

John Sackett has had trouble with a hoarse voice for the past four years. Whenever he uses his voice for more than two hours he tends to lose it. His family doctor corrected a problem with hypothyroidism thinking there might be a relationship, but there was no concurrent change in his voice. He has been accepted to law school and feels that he now must deal with his voice problem to be successful.

On endoscopic exam he has a white, smooth mass within his right vocal cord. When he closes his vocal cords together in order to produce sound, the plugged mucous gland prevents closure

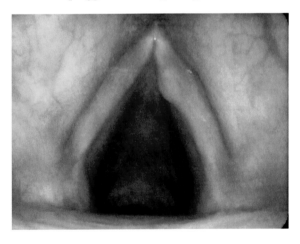

A white oval-shaped mass within the right vocal cord.

There are mucous glands along the inferior vibratory margin of the vocal cords that can become obstructed. When plugged (behaving similar to a pimple or sebaceous cyst on the skin), a cyst containing mucus forms. These occur anywhere along the length of the vocal cord, unlike nodules and polyps which are almost always in the center. When there is a one-sided swelling that is not in the mid-portion of the vocal cords, a cyst statistically becomes the most likely diagnosis.

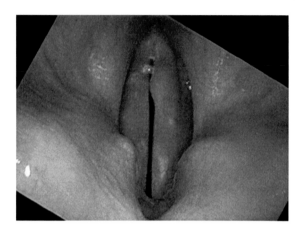

Stroboscopic view of the vocal cords in an adducted position. The right vocal cyst is protruding, creating gaps.

As a cyst swells, it impairs the voice by creating stiffness as well as a bump and so can impair the voice much like a polyp. If the cyst enlarges, the vocal cord vibrates at a lower pitch because of the heavier mass. Consequently, an examiner might hear husky hoarseness as well as diplophonia or roughness.

Occasionally, a cyst ruptures and drains out and sometimes they plug up again. Most commonly, cysts persist unless they are surgically removed. John ended up having his cyst surgically removed, recovering almost all of his vocal range. Because a cyst is deeper within the vocal cord then a nodule or a polyp, there is more often some leftover stiffness from scarring when the cyst is removed. His vocal edge was straight. His speaking voice was normal, but just there was just enough stiffness at the surgical site of the cyst removal to impair some of his highest pitches.

Saccular cyst

Barry Black is in his mid-20s and notes that speaking has gradually become effortful for the past 2 years. He feels like something is swollen in his throat although it does not affect his eating. He feels discomfort after speaking loudly for a period of time. His voice cracks like it did in puberty. He saw an ENT who told him that he had vocal cord polyps.

During his audible examination, his voice leaked a lot of air in his upper range so that he had a husky hoarseness when trying to sing in falsetto. On the endoscopic examination 2 large hemispherical masses were present on the bottom edge of the false vocal cords. When he closed his vocal cords together to produce sound, these put pressure on the true vocal cords and kept them from vibrating well or completely. The harder he squeezed, the more they rubbed against the true vocal cords.

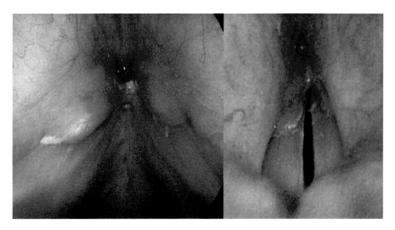

Left: *Saccular cysts protruding from the front and bottom edge of the false vocal cords in an abducted position.*
Right: *The saccular cysts are large enough that they rub against the true vocal cords during sound production.*

Healthy saccules typically contain mucus and are fairly small. If they become plugged, they may enlarge to this size or larger and dampen the vibrations of the vocal cords. He had the saccules surgically removed and his elevated vocal effort went away.

Laryngocoele

Desnea Winston had smoked about two packs of cigarettes a day for 45 years. Her lung function had deteriorated to the point that she was always out of breath and needed to be on oxygen. About one month ago she began to have trouble with her voice becoming more effortful to produce. Swallowing also became very difficult.

On examination she breathed with pursed lips like many patients with COPD. On endoscopic examination she had a smooth surfaced mass on the left upper glottis that appeared cystic. It did not block off her breathing but when she closed her vocal cords to make a sound the mass completely covered her vocal cords, likely touching them and severely dampening their vibrations. This made sound production difficult.

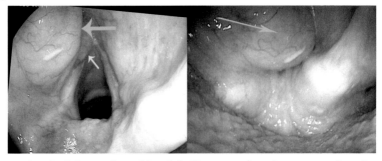

*Left: The left saccule is dilated (yellow arrow) and is connected to the larger sac or air filled pocket – laryngocoele (blue arrow). **Right:** during phonation the laryngocoele moves over the vocal cords covering them almost completely (thin blue arrow shows the direction of movement).*

During surgery, the mass was a cyst, mostly filled with air, but with some mucus present and was connected to the left saccule. She had filled the left saccule with air and apparently placed enough pressure on it to dilate and herniate the interior of the saccule up into the left false vocal cord. This is known as an internal laryngocoele.

During surgery the bulk of the cyst was removed and the remainder left open so that it would remain decompressed and could no longer expand, since I expect that her effort to breathe from her COPD will remain the same. These herniated pockets of air can also expand out of the larynx into the neck (external laryngocoele) and have been seen in horn players who have a great deal of back pressure on their larynx when playing a horn.

Thick mucus

"Doctor, the more I sing, the more I have to clear my throat."

I often see patients with sticky mucus, perhaps more often than I see patients with thin, translucent mucus. Sticky mucus clearly can alter vocal cord vibrations. However, I hear many people claim they drink copious amounts of water, yet still they have sticky mucus. Many people describe a sensation of a copious amount of post-nasal drip leading to throat clearing. Why do we have this mucus?

There are two types of salivary glands in the throat: serous and mucous. The serous glands produce secretions that are very watery. The mucous glands produce secretions that have a high concentration of proteins that makes mucus, which is much stickier than water. Perhaps there is some genetic predisposition to have thick or thin mucous. Maybe there is even some optimal ratio of the two types of secretions. Both types of secretions likely have roles to play. The thin secretions seem to lubricate the lining while the proteins from the mucous glands might keep the mucus adherent to the lining.

Personally, I notice that my vocal cord secretions are thicker and sticker after drinking an espresso. Diuretics, whether in the form of caffeine, alcohol or a pill (such as furosemide which people take for blood pressure) lead to stickier mucus on the vocal cords.

There are also drugs which seem to have an impact on the consistency of secretions. Guiafenesin is one of the most common drugs used as an expectorant, a mucous stimulant and thinner. You can find it in many cold medications. As I will mention later in a chapter on reflux ("Reflux laryngitis" on page 283), anti-reflux medication also seems to thin mucous out, at least in some people, perhaps increasing the ratio of serous to mucinous secretions.

More commonly than actual thick mucus, I find that secretions tend to accumulate anywhere on the vocal cords where vibrations are impaired. For example, a person with a polyp, nodule or other swelling will tend to accumulate mucus on the polyp while they are making sound and have to constantly clear it off the vocal cords. People with weak, thin or bowed vocal cords will tend to have secretion accumulation at the ends of the vocal cords where they come together. If the impairment of vibration is corrected, fewer secretions will accumulate.

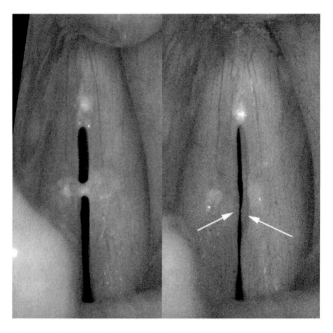

For the dedicated examiner, secretions quite often act as a divining rod, pointing to where a lesion will be found, rather than hiding it. There is a swelling on the both vocal cords, better visualized beneath the mucus accumulation after clearing. The mucus accumulated in this region because of the dampened vibrations.

Trauma

Injuries of force to the larynx

Internal

Intubation

Ms. Julia Achen woke up after general anesthesia for back surgery with a very sore throat. The discomfort persisted for about three months. Her voice was also hoarse immediately upon awakening. Her vocal quality still remains poor. Some days her volume is reduced to a whisper. She can no longer sing in her upper range. Her voice gives out or it cuts in and out during use. She can no longer yell loudly.

When she first inquired about her voice problems, her surgeon said, "Given the size of your mouth, it might've been a difficult intubation, but most times these problems resolve on their own." The surgeon was assuming that the anesthesiologist somehow rubbed the vocal cords when putting the endotracheal tube in, perhaps causing swelling.

She did not improve and eventually, she saw an ENT. Initially he suggested that her problem might be from acid reflux. On a follow up exam, he thought one of the arytenoids may have been dislocated when the anesthesiologist was trying to insert the breathing tube[16].

When I saw her, the right vocal cord was much looser than the left, which allowed air to leak out the right side. It appeared to me that her right vocal cord had been torn during the intubation and had healed with less tension. There was also a small white deposit within the vocal cord, at the back edge.

---

16 Although "arytenoid dislocation" is a frequently suggested injury, it is an extremely unlikely injury. The arytenoid cartilage is likely to fracture at much lower forces than for the joint capsule to tear and allow the joint to dislocate.

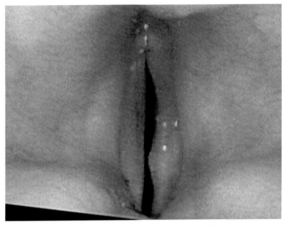

The right vocal cord appears much looser than the left.

I encourage anesthesiologists and surgeons to refer everyone with a hoarse voice after surgery, not only for the patient's benefit, but for the physician to learn what complication he may have had. My experience is that most physicians, including surgeons and anesthesiologists, have a tendency to tell people with a hoarse voice after surgery that they will get better. While most do, many individuals after surgery have some ongoing voice limitation and anesthesiologists and surgeons do not learn about the problems they cause when they ignore them, so their technique is never modified. Singers, who are basically vocal athletes, notice these types of limitations more than regular speakers. Although some injuries can be improved upon, other injures may not be correctable. The better care would be to learn how to avoid injuring the larynx. Seat belts, invented for the auto, result in better facial features for many people because suturing facial lacerations from the windshield is not as cosmetic as never going through the windshield.

Intubation granuloma

When asleep under anesthesia, the endotracheal tube puts pressure on the vocal cords near the vocal process. The cartilage is just below the surface and the mucosa overlying the cartilage is easily injured just from pressure. See "Intubation injury" on page 125.

Granulomas of healing

Granulomas basically can occur anywhere the mucosal lining has been left open to the bacteria of the throat. They may occur on a surgical site or on an overuse ulcer (see the story "Granuloma of overuse" on page 123 for comparison. See also the "granuloma photo" on page 222).

Stripping

We met Effie Forte earlier (see her story, "Effie Forte" on page 206). She was left with an asymmetric tension of her vocal cords after a vocal cord stripping. Stripping is becoming less common, but there are a number of surgeons still performing the procedure. Prior to the development of the delicate instruments, lasers and techniques to remove tiny bumps from the vocal cords that we now have available, there was a basic instrument called the cup forceps. Two sharp-edged hemispheres or cups that oppose each other would close together to bite off a piece of tissue. They were large enough to chomp off a piece of tumor as a sample for a pathologist to look at and tell the surgeon what kind of tumor was growing on the vocal cords.

As microscopes and endoscopes to visualize the vocal cords became more clear, smaller lesions like nodules and polyps became more apparent. Surgeons developed a technique where the cup forceps could be used to grab the nodule. Perhaps when brand new, the cup forceps would have cut through the nodule severing it off, but I never met a sharp pair. The technique evolved into grabbing the nodule in the forceps, twisting, then pulling, stripping or yanking the nodule off (gulp!). If you were lucky, just the nodule was removed. Far more often a long strip of mucosa ripped off. Too often the surgeon also grabbed too deeply and tore off not only the nodule and some of the surface tissue, but the lubricating layer of the vocal cord as well.

New skin grew over the injured area. If there was no nodule left and the edge healed straight, the voice could sound better. This apparent improvement resulted because this stripped cord was stiff enough now not to vibrate, so sound only came from the good cord. It is a bit like a person with double vision having one eye removed to improve the double vision. True the double vision is gone, but there is now not much reserve if there is a problem with the remaining good eye.

External

After a neck surgery

Johnny Argent fell down the stairs, suffered a brain injury and was in a coma. An emergency cricothyroidotomy was performed in order to place a tube that could be connected to a ventilator and breathe for him. The surgeon wanted to avoid stretching his neck, which could occur during a standard intubation, in case he had a cervical spinal cord injury. While most surgeons place a tracheostomy tube through the trachea a few rings below the larynx, Johnny's breathing tube was placed through the larynx between the cricoid and the thyroid cartilages, perhaps because it was an easier opening to find. However, this tube was left in place for two weeks, a long time to expose the cartilage of the larynx to the air and the bacteria from the skin edges. When he recovered from his coma, the cricothyroidotomy tube was removed and the opening healed over a few days. However, his voice remained soft and hoarse. He now would like to return to singing the blues, but his upper vocal range is severely restricted.

His uppermost pitch is A3, a note which is typically the top of a male's vocal *chest range*. He has no falsetto. On endoscopic examination, he cannot stretch his vocal cords, no matter how high a pitch he attempts to make and I can see that he is putting in a lot of effort.

A CAT scan of his larynx shows that bone has formed between his cricoid and thyroid cartilage. It is located at the edge of where the tracheostomy tube was placed and bridges the gap between the cricoid and thyroid cartilage. He can no longer contract the cricothyroid muscle since the two cartilages are fixed in place by this bridge of bone. The cricothyroid muscle is the one that puts us into falsetto and since he cannot move the cricothyroid joint, he has no falsetto. We also tighten this joint when we yell. Consequently, since Johnny's joint cannot move and lengthen his vocal cords, he cannot yell as loud.

He elected to have a surgery where the bridge of bone was removed and the cricothyroid joint was mobilized. After surgery, he had restoration of a small portion of his falsetto. He could get louder. However, the joint could not move as well as it did before his cricothyroidotomy. Most surgeons would avoid placing a cricothyroidotomy or if they needed to in an emergency, they would change it to a regular tracheostomy at the earliest time to avoid this type of later complication.

Surgery in the neck may uncommonly injure the structure of the larynx. A bit more common, blunt trauma from outside the neck may alter the structure of the cartilages of the larynx as in this next story.

Hit in the neck

Paul's grandson, John Bunyan, was working in his orchard when a log fell on his chest and neck. He arrived at the emergency room in a coma, where they placed a tracheostomy tube into his trachea to breathe for him. A fracture of his larynx was repaired, but when he woke, he had no voice at all. The first surgery was revised with a new plate on his thyroid cartilage and after this, some voice returned. Six months later, when he visits me, his voice is still lower in pitch, rougher in quality and softer in volume than it used to be.

On endoscopy examination, his left vocal cord is shorter than his right. The left vocal process is in front of the right vocal process when he closes his vocal cords.

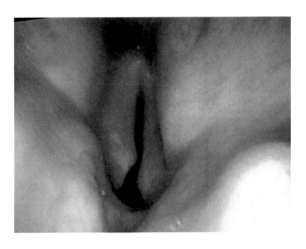

The left vocal cord is shorter and has less tension than the right.

This loosening of his left vocal cord (and possibly his right to a lesser extent) can explain all of his symptoms. With loose vocal cords, they will vibrate at a lower pitch than before the accident. With the left vocal cord shorter and looser than the right, he has a double pitch and he also leaks air between the gap in his cords. So, he has both a rough and a husky hoarseness. He cannot hold enough air below his cords to get as loud as he used to when he tries to yell.

With another surgery, I removed the plate and cut and length-ened his left thyroid cartilage. This stretched the left vocal cord. Though he was not all the way back to his pre-injury voice, his voice was stronger, smoother and higher pitched after this surgery.

Unusual disorders

Endocrine

Using testosterone

Lee B. Doe is nearing 50. For the past two years she has noted a cracking in her voice when she tries to sing in her upper range. During the interview I find out that she had been using testosterone cream to improve her sexual energy. When I listen to her voice, she develops a break in her voice at D5, then really cannot reach much higher. Most females can produce a sound at least a half octave higher than this when singing. On stroboscopy, at D5 her vocal cords are stiff enough to stop vibrating at low airflow. This is an atypical amount of stiffness.

Testosterone administered to a female seems to have the same effect that puberty has on a young male. It starts to lower the speaking pitch, perhaps by thickening the vocal cord and also slightly stiffening the vocal cords, gradually impairing the uppermost notes.

This can be put to good use in a transgender female to male patient. Over one to two years he will go through a transition in his voice just like puberty. However, if you are a female singer, administering testosterone can markedly impair the upper singing range and probably should not be used if one intends to keep singing in the upper range. While the changes are not reversible even if the medication is stopped, it is possible to retighten the vocal cords with a laser and gain back some of the pitch range.

Hematologic

Amyloidosis

Christina Amarillo is 30 and has been getting progressively more hoarse for the past 10 years. Her mother and three aunts are all hoarse. One of her aunts had surgery and amyloid was removed from her vocal cords.

Listening to Christina's husky voice, she leaks air. On endoscopy there are patches of yellow material underneath the mucosa of the vocal cords. These vocal cords are stiff and do not come completely together because of the yellow deposits.

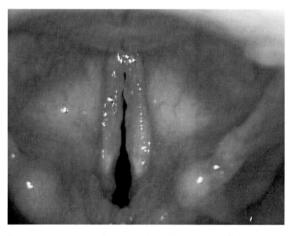

The amyloid is yellow in color and deposits mostly along the edges of the vocal cord, although there are some deposits within the aryepiglottic folds as well. The deposits on the vocal folds create a rough edge that allows air to leak out and they also make the vocal cords stiffer. Standard definition photo.

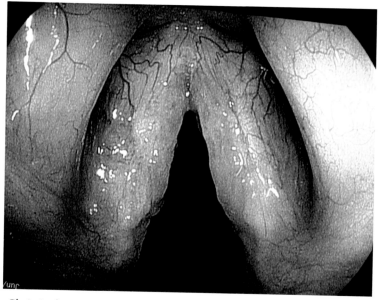

Christina's cousin also has amyloid deposits on her vocal cords. High definition photo.

In this condition, blood vessels seem to break easily and rather than the body reabsorbing all the blood, a yellow material is left as a deposit where the bleeding occurred. Deposits on the edge of the vocal cord create an irregular edge such that air leaks between the cords (huskiness) and the deposits create differences in stiffness and mass between the cords so that they vibrate at different pitches at times (roughness). Eventually more effort is required to start them vibrating.

Another type of amyloidosis consists of deposits into the false vocal cords. If the false cords enlarge enough with the deposits, they eventually can put pressure on the true vocal cords and dampen the vibrations.

The current treatment is to remove the deposits with surgery, very gentle surgery if the deposits are on the true vocal cord. The deposits will likely gradually reoccur over time. They can be removed again.

Autoimmune

Arthritis of the laryngeal joints

John Davenport was a music major in college and now pastors a small church. He notices that for the past two years he can no longer reach his typical upper notes. When he tries to sing he develops tightness in his left neck as if he is straining.

When I listen to him glide from his lowest note to his highest note that he can reach, he has only a one octave range. His uppermost note is C4 or middle C on the piano. This is an incredibly small range, especially for a trained singer. His upper notes have a tight quality. Basically he is unable to reach any falsetto.

During the endoscopic examination, I asked him to glide up from his lowest note to his highest note and I notice that his vocal cords do not change length. Something is preventing his CT muscle from working. Either the SLNs which supply the CT muscle are not working or the CT joint is frozen in position. The joint can be viewed with a CAT scan, which I order

He has developed arthritis of his cricothyroid joint. Consequently when he tries to contract his CT muscle, the frozen joint prevents movement and stretching of the vocal cords. This also explains the discomfort in his neck on the left side as these muscles try to work overtime. We discuss, but don't pursue, trying to mobilize the joint surgically.

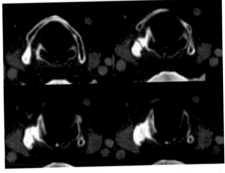

Everything in the x-ray should be symmetric. The large white area on the left side of the CAT scan images is the arthritis in his joint. The right side joint is normal.

Foreign body

Trapped

"I first became hoarse about five to seven years ago. It is hard to really remember when it first started since it came on gradually. About two years ago, I really noted the hoarseness increasing. It takes a lot of effort to speak and sometimes my voice breaks like a teenage boy. With even the slightest cold, I lose my voice completely. On the phone, people are always asking me if I smoke!" Julie Trappe reports, sounding concerned.

Listening to her voice, it is very rough and strained and she essentially has only a single pitch with an occasional higher pitched squeak. She has a maximum phonation time of seven seconds on one breath, quite short for a healthy person in her mid-forties.

On endoscopy there is a large, smooth surfaced swelling involving the back half of the left false vocal cord and also either covering or involving the top back portion of the true vocal cord. This is an unusual location for a mucocoele which is usually in the ventricle, between the true and false vocal cords, but near the front of the vocal cords rather than the back (posterior). Otherwise, the swelling looks like it could be a tumor. A CAT scan shows that the center of the mass is calcified.

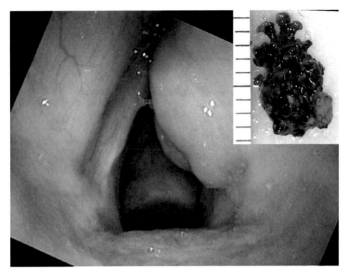

Swelling of the right true and false vocal cords is from a trapped "stone" (inset - scale is in mm) in the right ventricle between these cords.

At surgery, the tissue removed was inflammatory tissue and lodged in the posterior ventricle was an irregularly shaped stone or calcification. My suspicion is that a food particle became trapped in the ventricle between the false and true vocal cords and the secretions from the throat gradually deposited minerals such as calcium on the particle. Over many years as the deposits gradually increased the size of the stone, the inflammation increased. A swollen vocal cord doesn't vibrate well and a swollen false vocal cord that rubs on the true vocal cord impairs vibration further. She healed up in about two months with no further problems.

It is possible for foreign bodies to become lodged between the vocal cords. Some foreign bodies are small enough to pass between the vocal cords and lodge in the lungs. Small bones from fish or other meat products can impale themselves. Surgical implants in the vocal cords can extrude and lodge in the larynx. Almost any foreign body of the appropriate size could lodge in this area.

Combinations of disorders

Complexity

Finding the primary source of hoarseness

Verna Solo is almost 80. She noted the onset of hoarseness for three months but another problem took precedence. Her aorta had been enlarging for several years and her cardiac surgeon felt that she needed surgery to repair her dilated aorta last month. She woke up from surgery with only a whisper. While her voice has returned, it remains very raspy.

In thinking about possible reasons for her hoarseness, one consideration is that she could have a weak voice because the dilating aorta stretched her recurrent laryngeal nerve, causing a nerve injury. So perhaps her progressive hoarseness before the surgery and the enlarging blood vessel are related. Her weak voice immediately after a surgery near the recurrent laryngeal nerve suggests that a nerve injury during surgery is a possibility to be considered as well, and we might find a paralyzed vocal cord during endoscopy. We also know that endotracheal tubes in the throat can cause an injury and weakness after surgery.

Indeed, I find one of these possible explanations when I look with the endoscope: her left vocal cord is not moving well. It is not opening and closing very much during breathing nor during voicing. However, it does sit near the midline.

Sorting out her case requires thinking about a few complexities. She can almost close the right vocal cord against the left, so she has already compensated quite a bit for the paralysis. On endoscopic exam it is very difficult to see the left vocal cord. Besides nerve injuries, the endotracheal tube used during surgery can cause trauma. Although not common, a person may have more than one problem at one time.

I ultimately numb her vocal cords so that I can place my endoscope underneath the smooth, round, swollen left false vocal cord.

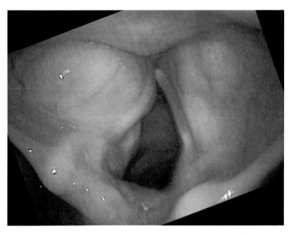

The left false vocal cord is dilated and obscures the left true vocal cord. There is a fluid filled cyst enlarging the left false vocal cord. The left vocal cord is also not moving and in the view during inspiration, the left vocal cord remains positioned near the midline.

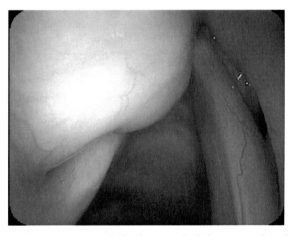

Even on a close-up view it is difficult to see the left true vocal cord. I actually have to lift the swollen false vocal cord up with the endoscope to see the true vocal cord.

The saccule is a gland located on the bottom of the false vocal cord at the front of the larynx. Usually the saccule cannot be seen from above, rather it is visible when the endoscope is close to the larynx and looking parallel along the vocal cords. It then appears as a small hemisphere along the bottom of the false cord within the laryngeal ventricle. However, occasionally this gland may plug and, as in Mrs. Solo's case fluid accumulates inside the gland. The saccule swells with its own fluid creating a saccular cyst.

Now we must consider that we have a complex problem; a left vocal cord that is not completely paralyzed – it moves some. So there is synkinesis of the left vocal cord. The right vocal cord closure has compensated to the point that she can bring the two vocal cords together. Consequently, even though the left cord's neurologic injury is visually obvious, since the right vocal cord closes against the left during sound production, they are not causing air leak (husky hoarseness). Since the left cord has reinnervated, it also has the same bulk as the right vocal cord and so it is not causing diplophonia (rough hoarseness).

I anesthetize her vocal cords and put the endoscope almost on the vocal cords and watch them vibrate when she makes sound. The left-sided saccule, has enlarged to the point that it pushes down on the true vocal cords. This pressure on the vocal cords dampens their vibrations. It then requires more air or effort to start and to keep the vocal cords vibrating. Her saccular cyst is actually causing her vocal effort and her weak voice by rubbing on both vocal cords when she tries to make a sound. Even though she has an obvious left-sided vocal cord movement impairment, it is only a striking visible abnormality and not actually the cause of her weak voice.

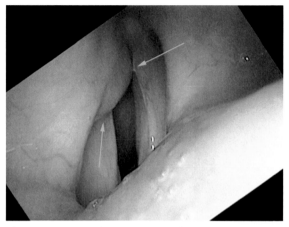

During vibration the saccular cyst touches both vocal cords. This photo is taken during the most open phase of vibration. The cyst touches the right cord on the vibrating edge (right arrow). The cyst rides continuously on the left vocal cord, touching beneath and out of our view in this photo.

As an examiner, it is sometimes possible to become so focused on finding an expected problem that something large and in the field of view may be missed because of the examiner's strong expectations. I have seen cases almost identical to this one, where the ENT entirely missed this large saccular swelling because his mental expectation was to find a problem on the vocal cords themselves. It is even more unfortunate when an examiner misses a cancer of the false vocal cord because they are so focused on the true vocal cord. In the end, as Mrs. Solo lived alone, she decided not to have a biopsy, an excision or anything done surgically to improve her voice.

When a person comes in with a complaint of, "I am hoarse." This should not be interpreted as, "Please find all the abnormalities in my throat and fix them," but rather, "Please find the problem that is causing my hoarseness, tell me what can be done about it and what are the risks." In the context of hoarseness, this means that we must find the gap or the asymmetry or both depending on what type of hoarseness they have.

Part III – Examination

The primary care exam

What can a primary care doctor infer about hoarseness?

If we recall Faith N. Metsan's doctor's visit in Part I, we recall that she lost her voice suddenly and she went in to her primary care physician's office (Marcus Goodew) for advice. He performed a number of typical medical tests and looked into her mouth, but as best we can tell, he never looked at her larynx nor her vocal cords. On what basis did he make a diagnosis and on what basis did he offer her treatment advice?

Dr. Goodew's evaluation must have been largely based on a hunch or a guess, since there is no evidence that he ever saw the vocal cords. That guess might have been based on some sense of whether the hoarseness could be due to a common transient illness – a cold. Perhaps colds have been common for a few weeks in his office. Perhaps some symptoms Faith described matched a cold.

Let's consider another patient similar to one we have seen before. Amelia Desire reports annual hoarseness for 30 years that typically starts up in both April and in November. It lasts two to five weeks and often goes away on its own. Her voice fades out as she uses it and she reports discomfort in her neck if she speaks very long. She rates herself as a 4 on the 7 point talkativeness scale. Many times she sees her primary care physician and reports that two courses of antibiotics often seem to make the hoarseness resolve. However this year, the hoarseness persisted after two different antibiotics. Her PCP then tried a course of steroids "to reduce the swelling," but her voice did not improve. He then tried a one month course of a proton-pump inhibitor since she does have a sense that acid comes up into the back of her throat. This treatment helped the reduce the sensation of acid in her throat, but her hoarseness persisted. Then, since her hoarseness seems to often coincide with the springtime allergy season, he tried a steroid nasal spray and an antihistamine. Her voice remained hoarse and she was referred to me for an evaluation.

I listen to Amelia and when she speaks, her speaking voice is at a very low pitch for a female yet occasionally breaks into a high pitch. When she laughs, her voice is normal. When I ask her to yell, she hesitates at this vocal task. When I have her sing a high pitch, her voice is very clear and she smiles, somewhat inappropriately. I examine Amelia's vocal cords with an endoscope and her vocal cords are normal in appearance, though she is often holding them apart during phonation. She happens to be speaking with her thyroarytenoid muscles relaxed and her surrounding neck muscles very tight. She has a nonorganic hoarseness. She can speak with a clear voice at the higher pitches, so I start there and then gradually move her speaking pitch lower until she has a clear voice at her normal speaking pitch. Within a half hour she can even change from a hoarse voice to a clear voice on command. Her hoarseness has resolved in the office.

Since her husband is present during the exam, I discuss how muscle tightness causes her voice problem and how stress often aggravates muscle tightness and plays a role in the problem. I also discuss how the muscle tightness becomes a habit and that the voice problem is quite likely to return again in the future, especially if there is any stress in her life (she volunteers that she has plenty of stress in her life.) Stress frequently causes muscle tension. Finally, I focus on how this disorder could spontaneously resolve and how the assumption that it went away in the past with an antibiotic was an erroneous interpretation of coincidence representing causation. I caution that two courses of antibiotics twice a year, steroids, anti-reflux medication and anti-allergy medication are not only a consumption of her time and her money, but over many years, she is at risk for antibiotic resistance, fungal infections or other medication side effects. I had her work with a voice therapist to solidify the gains she has made today in the office.

Now, let's dissect the approach already taken to hoarseness in Amelia.

Bacterial infections of the larynx are rather uncommon. When there is a bacterial infection of the larynx, it often causes rather severe swelling and, often enough, even airway narrowing as well as rather significant pain. Antibiotic treatment of hoarseness is quite low on my list of treatment options for laryngitis, that I have probably prescribed antibiotics for a vocal cord problem less than 5 times in 9,000 patient visits. I would move this treatment paradigm to low on my list.

What about steroids? Steroids do reduce swelling very effectively. Typically prednisone or methylprednisolone taken as a pill will reduce swelling on the vocal cord within two hours. If a patient has swollen vocal cords (and a corresponding deep pitch from that swelling) these symptoms can be temporarily alleviated with steroids.

The steroid effect of reducing swelling is typically to reduce edema or fluid build up, but is otherwise fairly non-specific. Whether the swelling is from a general enlargement and stiffening of the vocal cord, such as from a viral infection, or whether it is from a discrete swelling along the margin of the vocal cord, such as a nodule and overuse, steroids will reduce the swelling. The improvement in the voice is rapid, but temporary. In this manner, treatment with a steroid can almost be used as a test. If Faith's voice improves within a day of taking a steroid and her pitch rises, her hoarseness is likely due to swelling.

In practice, Dr. Goodew has heard from his otolaryngology friends and colleagues that reflux is the main cause of hoarseness, so he prescribes an anti-reflux medication (occasionally the newest and most expensive one). Since nearly every prior referral to an otolaryngologist of a patient with a complaint of hoarseness came back to his office with a pill and a diagnosis of reflux, Dr. Goodew (not unjustifiably) hopes to preempt the wait to get in to the otolaryngologist and prescribes an anti-reflux medication from the start of the symptoms. As for this reflex to treat reflux, with proton-pump inhibitors or other anti-reflux measures, I will defer a more complete discussion (see "Reflux laryngitis" on page 283), but I would encourage Dr. Goodew to hold onto his pen for now before putting ink on the script.

Do allergies cause hoarseness? Presumably there could be an histamine reaction along the edge of the vocal cord causing swelling similar to what happens so often in the nose. Yet the lining of the membranous vocal cord is a very different epithelium than the lining of the nasal turbinates. If there is edema of the vocal cord from allergies, presumably the patient would be complaining of a lowered pitch. Given that the nose is the principle filter of air entering the lungs, and the large number of people experiencing nasal symptoms from allergens without a corresponding change in their vocal pitch, I again suspect that allergen effect on the vocal cords is minimal or infrequent, at least relative to the nose. While allergy treatment is relatively benign – many allergy medications are available over the counter – patients are paying for Dr. Goodew's judgement, so I would cautiously put allergy treatment low on my differential diagnosis.

Using a pulmonary steroid inhaler with the thought that steroids applied topically to the larynx will reduce assumed allergic laryngitis is fraught with an increased incidence of hoarseness as these patients may develop a fungal infection on the vocal cords, stiffening and thickening the vocal cords from the fungal growth. There is no reason to want to participate in iatrogenic hoarseness. I would save inhaled steroids for patients with asthma and even then I would use the lowest possible maintenance dose to avoid giving them a hoarse voice from fungal growth on the vocal cords. In the past 10 years the incidence of fungal laryngitis from steroids has become more and more frequent.

If Dr. Goodew should keep all these potential medical treatments low on his list of things to do for hoarseness, what can he do to be helpful to his patients? Let's think about how he could improve his hunches.

Dr. Goodew should first ask himself if he is hearing a speech or a voice problem. If the patient has a difficult time forming words, the problem is likely coming from within the mouth – above the speech line. I would think first about neurologic problems that could be impairing motion of the lips, tongue and palate or conditions that involve the area of the brain that controls these muscles.

Then ask, is the problem a *quality of sound* or *lack of volume* issue. Is this either a husky voice or a rough voice or a mixture of both? Is this a weak voice? If the answer is yes to any of these, the problem lies on the vocal cords.

I would next determine if the hoarseness likely belongs to one of the behavioral hoarseness categories. Is the patient either a vocal overdoer (they give a high score on the 7-point talkativeness scale) or a vocal underdoer? A patient with hoarseness who rates herself a 6 or a 7 on the 7-point scale has at least an 80 percent probability that her hoarseness is due to a swelling on the edge of the vocal cord just based on this information alone.

Did the patient lose her voice rather immediately after loud vocal use and then suffer from sustained hoarseness? Think of the likely probability of a hemorrhage or other vocal cord edge swelling. Patients with these disorders are quite likely to experience a temporary improvement with steroids. I seldom use steroids for these disorders as I feel the hoarseness responds best to long term behavioral changes. Steroids are more like a crutch to get the patient through some temporary vocal requirement such as an upcoming performance or presentation. However, if you use steroids and the voice improves rapidly, then some type of swelling is likely the underlying cause of the hoarseness.

Is the patient an underdoer and complaining of hoarseness as well as neck discomfort with episodes of prolonged vocal use? Think about bowing or muscle atrophy that is often accompanied by neck pain from neck muscle tightness. People with thin vocal cords compensate by using the accessory muscles in the neck and, in general, bilateral neck pain with voice use is a very frequent complaint.

Ask the patient to make a few different types of sounds – low pitch, high pitch, loud and soft. Is there an inconsistency where some sounds are clear and others very rough? Does the patient have no sound production at all, but appears to be holding back or not putting effort in toward making a sound? Think about nonorganic voice disorders or inappropriate vocal muscle use.

Behavioral voice disorders such as overuse, underuse or inappropriate use are very common. If the hoarseness resolves quickly – days to a week or two – no further treatment is needed unless the patient desires a more accurate diagnosis.

Next in the line of query, did the patient just have a surgery? Injuries to the nerves supplying the larynx are the most frequent disorder I see in my practice. Surgery in the neck or in the chest can injure the vocal nerve supply directly. Thyroid surgery, parathyroid surgery, anterior cervical fusions, carotid surgery, lung surgery or heart or aortic vessel surgery are all frequent culprits of recurrent laryngeal nerve injury. Even the breathing tube placed in the lungs during any general anesthetic surgery can put pressure on the nerve from inside the airway and cause a vocal paresis.

While "watch and wait" for spontaneous recovery is the most frequent paradigm in use for nerve disorders, I find that intervention to immediately restore the voice is often appreciated by patients, especially those who are innately talkative or whose voice is important for work. So if the history is suggestive of a possible nerve injury, I would send the patient on to a laryngologist. Let the patient make an informed decision on whether early treatment is appropriate for her amount of voice use.

Voice disorders that come on gradually, have none of the above more obvious explanations for their etiology, and/or persist for more than a week or two probably warrant an examination to determine their etiology - structural or behavioral. I really know no other way to diagnose tumors, fungal infections or paresis without a visual examination of the vocal cords. If Faith has a need to know the cause of her hoarseness, she has a need for an endoscopic examination.

However, both Dr. Goodew, as well as Faith, the patient, can utilize the concepts in this book to take an educated guess at the probable cause of hoarseness and determine whether the expense of a laryngeal examination is warranted. Each can weigh the cost of being off work and the cost and risks of inappropriate treatment and delayed accurate diagnosis against the cost of a thorough endoscopic examination. My hope is that Dr. Goodew tells his patient that since he cannot see the vocal cords, his diagnosis is an educated guess, and if the patient needs a more precise diagnosis or fails to improve in a brief period of time, then a referral to a laryngologist would be appropriate.

Seeing the vocal cords

When you get to match what you see with what you hear

Mirror examination

The laryngeal mirror placed in the back of the throat and used with a headlight has been around over a century. There was clearly a time when this exam was state of the art and there are places where it likely remains state of the art. Clearly it is an art to hold a mirror steady in the back of the throat, visualizing a reverse image of the vocal cords – many times while the patient is gagging.

Certainly, an examiner with patience can see the vocal cords. Moderate to large lesions such as tumors are visible. However, the examination is not recorded for later slow motion review. Indeed, much of the vocal cord motion is hidden by either the speed with which it takes place or by the visual limitations of using a mirror where portions of the vocal cord are obstructed by other structures.

If that is all that is available, it is better than a look into the mouth with a light.

Endoscopic examination with the naked eye

The next level of examination is using an endoscope, either through the mouth or through the nose. The examiner places his eye up to the endoscope to see the vocal cords. These devices help bypass some of the anatomic obstructions – the tongue, the palate and the gagging. Still, without a recording that can be reviewed in slow motion, it is likely that many causes of hoarseness will be missed. Additionally, without a recording, it is impossible to go back in time and review and learn from inaccurate judgement.

Complete & adequate voice exam

History

Almost every type of physician-patient interaction begins with a history of the problem. The physician says, "Tell me what happened." The story the patient tells about her problem is the history. In laryngology, three things are accomplished simultaneously during the history. This history offers a great many clues about likely types of voice disorders based on when the hoarseness started, simultaneous events at the onset, the duration, the severity, changes with use and simultaneous breathing or swallowing problems. Secondly, the examiner gets to begin hearing the conversational voice for a period of time. Third, the patient highlights preconceived perceptions of her problem that are also important for the physician to address later in the exam.

People who talk a lot will suffer from a different set of problems than people who are naturally quiet. People who lose their voice suddenly after yelling will have a different problem from those who lose it slowly over time.

The history is the first of a three part examination:

1. history,

2. vocal capabilities testing and

3. visual exam.

A three part examination has the same value as navigation at sea by triangulation. With only one point of reference, the sailor will know only the bearing or direction toward something, and with two points of reference she might be able to narrow her location on the map to a few possibilities, but with three references, the sailor can be almost certain of her exact location. Likewise, when all three portions of the laryngeal exam point to the same problem, the physician can say with great assurance, "This is the likely diagnosis."

My simplest first step is to eliminate speech problems. If the volume is loud enough and clear enough that people can hear the patient, but still not understand the words, the problem is likely a speech issue. Problems with language are speech issues. Verbal rate and articulation issues are speech issues.

For example, a person with a cleft palate cannot close the palate completely against the back of the throat. Closing the palate completely against the back is required to make certain consonant sounds, otherwise air leaks out the nose and those sounds cannot be made. /p/, /t/, /k/ as well as /s/ and /sh/ are typical sounds requiring complete closure of the palate, channeling all of your air out the mouth. If you try to say "pa, pa, pa" with the palate relaxed, air leaks out the nose and you sound to others like you have a cleft palate. From a physician standpoint, when I hear this nasal air leakage, I know to look at the palate. I really do not even need to look at the vocal cords. So hearing the problem directs me where to look.

If the problem is with consonants, vowels, words and sentences, the problem is likely not with the vocal cords. Voice issues should usually involve complaints about pitch, volume or clarity. Patients may not actually use these words. Volume and roughness qualities are voice issues that non-singers might typically complain of and pitch, clarity and register are issues singers may more likely notice.

A person might say, in addition to "I am hoarse," something like, "I can't get loud enough to be heard in a restaurant," (volume problem) or "I sound like I have a frog in my throat" (clarity problem). A singer might say, "I am missing a few notes" (pitch issue).

Vocal capabilities

Hearing the voice problem

Having determined that the problem is with the voice, how do we determine the specific problem and then the cause of that problem? If the vocal cords alter pitch, volume, clarity and register, we can strive to set all of these parameters constant, then by varying only one at time, we will most efficiently find our problem.

Generally, the patient doesn't even need to speak words during an examination of a vocal problem. Singing is far more fruitful for finding a vocal problem than speaking. Perhaps I should say, making various continuous vowel sounds is most helpful, as some people feel they cannot sing. If the examiner can listen to a sound at various volumes and various pitches, a vocal impairment will be more easily elicited.

Robert Bastian has described this process as vocal capabilities testing. By modifying one parameter at a time, when we hear the problem, we will know quite likely where the problem lies, even before we look at the voice box. Then when we look, we will not be distracted by the color of the vocal cords, for instance. We will go straight to where the problem should be and find it.

The vowel /i/

I prefer the vowel sound "ee" for my exam. In phonetic English I would write /i/ (or /i:/) as in the word f<u>ee</u>l or b<u>ea</u>d. This vowel utilizes the most upright position of the larynx and most open position of the pharynx or throat above the voice box. It makes examination of the vocal cords with an endoscope easier. The /u/ sound is a close second, as in wh<u>o</u> or b<u>oo</u>. The other vowels /æ/, /e/, /o/ tend to move the tongue and epiglottis back, narrowing the throat and make visualization of the vocal cords more difficult.

Remember, simplify! I pick one vowel and generally stick with it during my exam, so there is one less variable to deal with. Then I put the voice through a series of tests with this vowel.

Why a series of tests?

A familiar medical analogy comes from cardiology. A person complains of chest pain and an EKG test is performed. Let's say the result appears normal. Rather than telling the patient "you are normal," the cardiologist continues with a stress test, having the patient run on a tread mill while still hooked up to the EKG, perhaps even to the point of reproducing the chest pain during the test. Now the EKG appears abnormal in association with the elicited pain. Many more problems will be found by stressing the system, whether it is the heart pumping blood or the vocal cords making sound.

If the patient just says /i/, there may be no obvious hoarseness at one particular pitch or one particular volume, but it may appear at another pitch or volume. So stress test the voice.

Of course, for many people just showing up at the doctors office is stressful. Then the thought of singing to the doctor elevates stress to a new height. But that is not the stress to which I refer.

Reading

I record the patient's voice by putting on a headset microphone held out in front of the mouth. I have the patient read a paragraph from a book out loud in a comfortable voice. I use the same paragraph for every exam. It may be boring, but I record the same passage on every exam so when the patient returns in the future, it is very easy to compare. I know that reading is a mixture of voice and speech, but reading does several things for me.

Reading relaxes the patient and takes the focus away from the examination – most patients start out with a great deal of anxiety about what foul tasting medicine I am going to put in her throat, how big the tube is that I am going to put in her nose and how much it will hurt. These fears are not irrational as many patients who have come to me have been examined previously and the medicine did taste terrible, the exam was very uncomfortable or even painful and they may have gagged terribly.

While I have not solved the taste issue, only ameliorated it, the pain issue should be non-existent. We have such excellent topical anesthetics, that the most a patient should experience is a light pressure during an exam with an endoscope. For the terrible gagger, with a combination of topical anesthesia and the patience to allow the anesthetic time to work, there is no one I have not been able to examine.

Gagging can be completely psychological. I have patients who gag at the sight of an endoscope, but there are still techniques for getting around the mind. Anyway, starting out with a simple reading task puts most people at ease.

Second, by listening, I determine the approximate average speaking pitch by matching the voice with a note on the piano. It is not necessary to know the precise pitch, though there are machines that can do that. I am only looking for an approximation, indeed we typically modulate our comfortable speaking pitch over several notes to convey emotion. Good storytellers modulate a great deal but there will be an approximate central pitch. We use only a very small portion of our vocal range in daily speech.

Third, the reading task allows me to listen and determine for myself if there are any speech issues. Problems with the rate of speaking or poor enunciation become audible while reading. I may not yet hear the vocal issue or the hoarseness, but I can decide whether speech is a problem.

Maximum phonation time

Then I switch to the /i/ sound and ask the patient to see how long she can say /i/ on one breath at her comfortable speaking pitch and volume. I am not controlling pitch and volume as precisely as a researcher might with computerized testing equipment, but as with many things in life, we do not need to measure a 2x4 piece of wood to the nearest 1/64th of an inch in order to build a wall in a house that will function quite well for many years.

This test, the maximum phonation time (MPT) is a rough measure of how completely closed the vocal cords are. The more closed they are, the less air wasted and the longer the sound can be maintained. As a rough guide, with an MPT of less than 10 seconds duration most people will complain of being out of breath with talking. Healthy young people can typically go beyond 20 to 30 seconds. There are many variables which effect this test including lung capacity and vocal strategies used to produce the sound, but the more that the pitch and volume are kept constant, the more the test measures vocal cord approximation. This is especially true and helpful for one individual over time, since when I implement some treatment for the voice, any change in MPT after the intervention is likely from the intervention.

Vocal range

Next, I measure the lowest note and the highest note the person is capable of producing, at any volume, to define the pitch range of the voice. Sometimes the person has excellent vocal rapport. I can play the notes on a piano and the patient can match the note with her voice. Some people are not so talented and I will often ask them to slide up or down in pitch and then with my ear try to determine their lowest and highest note.

Swelling tests

After obtaining this maximum range, I try to assess the upper and lower ends of the vocal range at the very softest volume the patient can produce. Quite often, this requires some coaching. There are a number of disorders that impair soft voicing and despite the patient's interest in solving her problem, no one likes to "fail" at a test. This is especially pronounced in singers. Even when they complain to me that they are missing notes, they try their hardest to avoid sounding "bad" on those notes for me. I keep coaching them to go softer and softer and emphasize that I want to hear when the vocal cords stop vibrating or when their voice sounds bad as that is the purpose of their visit.

Generally, we should be able to produce the extreme upper and lower notes of our range at both loud and soft volumes. When we cannot reach the same note softly that we can reach loudly, there is probably a vocal impairment. The greater the difference in vocal pitch range, between loud and soft voicing, the more significant the problem.

One of the easiest ways to determine the upper soft range is to have the patient sing the first four words of the nearly universally known song, "Happy Birthday." When singing the words, "Happy Birthday to you, between the word "day" and "to" is a melodic interval of a fourth. If no sound comes out on the word "to," or if there is a significant onset delay to the start of vocal cord vibration on that word, then there is some mechanical change in the larynx within this interval of a fourth. This test can be repeated up or down a note and the point where the voice cuts out denotes the soft cutoff point.

Robert Bastian has termed this test for the soft, upper vocal ceiling, the "vocal swelling test" and is very diagnostic for nodules and polyps (see "Vocal overdoers" on page 89). In general, the point at which there is an onset delay signifies the point where a swelling on one vocal cord touches the other vocal cord and stops the vibration. It is just like putting your finger lightly on a guitar string, dampening or stopping the vibration.

It is also possible to learn to hear a central glottic gap with this test. The point at which the patient cannot start the vocal cords vibrating (because all the air leaks out between the cords) does not occur at as precise a pitch as when a swelling stops the vibrations. But there will be a general, but varying pitch where the vocal cords cannot be entrained because of the gap. This test tends to help me the most in pinning down a diagnosis, especially for behavioral voice problems.

Volume

I ask the patient to yell robustly, not a scream, but a well supported yell on the word "Hey." This additional stress from increased pressure beneath the vocal cords can cause weak vocal cords to flutter. The task may allow stiff vocal cords to actually produce sound, when quiet sounds were almost impossible. Psychogenic problems often show up on this test when the patient hesitates or exaggerates performing this task. Yelling is a simple task that everyone knows how to do. The important point is to note if the voice is different on this task than it was on the previous tasks.

Vegetative sounds

Lastly, I ask the patient to cough, followed by a clearing of her throat. This task can be helpful again in sorting out weakness of the glottis or psychogenic or nonorganic vocal problems. For instance, if a patient could only whisper up to this point in the exam, but can produce a robust cough, then the vocal cords have the capacity to come together and generate sound. They were likely being held apart with muscle tension up to this point in the exam.

Summary of vocal capabilities

I find the above vocal tasks sufficient to elucidate almost all vocal disorders. At a minimum, they will direct me where to look and what task to ask the patient to perform when I look with an endoscope or when I turn on the stroboscope. In fact, since I have been performing these tests often enough and long enough, I can hear and identify voice problems just walking through a crowd.

One of my hobbies is traveling and meeting new people. I try to hear the person's accent and their vocal character, and then at some point into our conversation, I can make some predictions about where they are from and what they are like socially. The accent of your speech and the character of your voice tell a lot about you, particularly if you are a member of the vocal overdoers or vocal underdoers of the world.

Laryngoscopy

Various types of endoscopes

At this point in the exam, the laryngologist should already be able to predict what he is going to find and a laryngoscopy is the key that reveals what is behind the door. The history and the vocal capabilities testing should suggest one diagnosis pretty strongly or have narrowed the problem to a few potential issues. What the physician sees with the endoscope should generally confirm the problem that his ears tell him is already present.

The word laryngoscopy can be broken down into two parts. *Laryngo* – the larynx or voice box and *scopy* – to look at.

The essence of laryngoscopy is getting a light to the larynx. As stated earlier, it can be as simple as putting a mirror in the back of the throat and shining a light on the mirror. Remember that just a light and a tongue blade are not enough to see the vocal cords. The brighter the light, the better the view, though for a detailed view in a voice lab, moving beyond the mirror is essential.

Technology has come a long way in the short space of my medical career, and glass rods, flexible fibers and video microchips on the tips of flexible tubes have opened up a new world for the examiner. High speed cameras, which I am not presently using, might aid in confirming or refuting some of the conceptions created with stroboscopy, though at present high speed cameras are rather tedious for daily use and remain primarily in research laboratories.

For the well-equipped, standard voice lab, I would consider the following equipment essential to varying degrees:

- Endoscope
    - Rigid endoscope coupled to a camera
    - Flexible fiberoptic endoscope coupled to a camera
- Light source

- Recording device
  - Computer with adequate memory for video recordings
  - or videotape (analog or digital)
- Stroboscope

In the high-end laryngology lab, I would have:

- A flexible videochip endoscope,

- A flexible high-definition videochip endoscope,

- A rigid transoral endoscope,

- A videochip endoscope with a working channel,

- A stroboscope,

- A digital recording device (computer with video capture software and a large RAID hard drive) and

- A high definition video screen.

I don't use the rigid endoscope frequently, but it still offers an iconic overhead view of the larynx. It offers the best and clearest view of how swellings on the vibrating margin impair vocal cord vibrations. So I use it most in vocal overdoers where I suspect a swelling on the edge of the vocal cord.

At the lowest end, I would have:

- A flexible fiberoptic endoscope with a camera and

- Any digital recorder.

I would give up a rigid endoscope and a stroboscope before giving up a video recorder in the voice lab. Without a recording, viewing the larynx with only the naked eye loses too much in the way of documentation and inability to review what happens so quickly in laryngology. With this simple lab, the very astute observer could even elucidate the pathologic process on video recording without a stroboscope, though it takes a great deal of effort and intuition.

There are certainly other options to setting up a voice lab which can produce excellent quality diagnostic images. For instance I have seen a lab use a rigid 70° endoscope and tip it all away into the larynx for close-up views. While not a common approach, this can lead to high-definition laryngeal images in the hands of a skilled and patient examiner.

Another alternative that some labs are following is to utilize only a high-speed video recorder and videokymography to view the actual vibrations of the vocal cords. This has some advantages for viewing lesions on the edges of the vocal cord but loses some capabilities for finding laryngeal lesions not on the vibrating edge of the vocal cords.

Recording device

The human eye is a notoriously poor recording device. Our memory quickly forgets important findings. We completely ignore things that we see when we feel they're not important (and later they might be the most important finding). Consequently a camera is the most valuable device after having any one endoscope.

Rigid endoscopy

I am aware of 70 and 90 degree Hopkins rod endoscopes. I prefer the 90 degree glass rod as opposed to the 70 degree, both for the ease of obtaining an image and for the comfort of the patient in positioning their head during the exam. While the 70 degree endoscope can reach closer to the larynx than the 90, for me the value in the rigid endoscope no longer lies in its degree of closeness to the larynx, but rather in the standardization and the clarity of the view. The flexible endoscope far more easily provides the close view. With the mouth open, there is an alteration and limitation to phonation because the position of the tongue becomes fixed in the examiner's hands. In the voice lab with enough money for a fiberoptic endoscope and a rigid endoscope, the rigid endoscope offers a clarity of view that is complimentary to the flexible endoscope's close and functional view. The rigid endoscope is poor at assessing much function other than changes along the medial margin of the vocal cords because it is limited to a single perspective that is nearly directly above the larynx. There is no other choice as to where to position the scope. It gives a fairly detailed view of the microvasculature as well. The main functional alteration to be made while viewing is to observe how the vibrations of the medial vocal margin alter with changes in pitch. The images are beautiful for presentations.

In the high-end lab, I use the rigid endoscope primarily in the patient where I suspect there is a mucosal lesion, otherwise I use the chip endoscope on nearly every exam. In the mid-range lab, I would often use both the flexible and rigid scopes in many patients to fully define and characterize much vocal pathology.

Flexible fiberoptic endoscopy

There are many flexible fiberoptic endoscopes that are relatively affordable by medical standards. Their flexibility allows a varied functional view of the larynx and a varied perspective which is under utilized by many laryngologists. However, detail is lost because of the fibers transmitting the image. With the image in sharp focus, pixilation by the optic fibers coupled with the pixels of a digital image create a moire effect that detracts from the image (imagine a view through an insect's eye). The three alternatives for a soft focus are defocusing the camera, defocusing the endoscope or electronic image smoothing algorithms within the camera processor. I find no difference between the three means for blurring the focus. None the less, even with a less-than-clear image, the key for the laryngologist is to move the endoscope beyond the nasopharynx, and to get close to the glottis. Closeness improves the image. After topical anesthesia, the endoscope can be placed into all the nooks and crevices of the larynx, touching the larynx if necessary without provoking a gag response or a choking spasm.

Closeness is important as it improves the image in three ways, effectively creating a high-definition image. First, any pathology fills more of the camera lens, and since it is a very wide angle lens, even a slight bit closer greatly enlarges a finding. With more pixels the pathology is clearer. Second, the closer the camera is to the structure, the more light placed on the structure and the brighter the image. Third, most processors have an auto-gain feature so that the image always appears with an appropriate brightness, typically as bright as possible. That means that when the endoscope is far away, rather than a dark image, the pathology is seemingly well lit. This appearance is due to increased gain or electronic light amplification. The trade-off for the brightness is more electronic noise. In the endoscopic image, this noise adds a lot of additional red dots within the image. The higher the gain, the poorer the precision of the image and the more "red" the image appears.

The examiner with a flexible endoscope who moves the scope closer to the vocal cords overcomes the three deficits to a great degree. The image is clearer, brighter and has less red artifact.

Flexible chip endoscopy

Image capture technology has moved the camera from an attachment to the eyepiece of the fiberoptic scope onto a chip at the tip of the endoscope, so that the image no longer needs to travel from the tip though fiberoptic cables to the camera. This technology transmits a very clear, non-pixelated (by fiberoptic tubules) image to the monitor. There clearly is some pixilation at some resolution with any digital technology. This view is really only compromised by the size deception of the wide-angle lens.

If the examiner never gets much closer to the larynx than the tip of the epiglottis, small pathology can still be missed even with this very expensive instrument. This limitation is bypassed by approaching the larynx. Getting close to the larynx takes skill in technique and patience, and more of the examiner's time. Essentially, in the mid-range lab, the rigid endoscope is the telephoto lens that allows a detailed view from far away, but that same view can also be improved upon with the videochip scope by approaching within a few millimeters of the vocal cords.

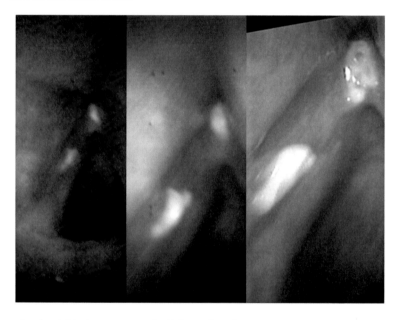

Leukoplakia is present on the left vocal cord.
Left: *image taken from a medium distance through a flexible fiber-optic endoscope. There is significant video noise and red speckling from the automatic video gain of the camera.*
Middle: *just moving closer improves the lighting and reduces the artifact from video gain although the moire pattern or soft focus will always be visible distorting the image from the fiberoptics.*
Right: *with a flexible chip endoscope his leukoplakia comes into a clearer view.*

Photos taken with even the best technology are strongly affected by how close the examiner comes to the vocal cords whether with fiber optic, chip or even high-definition endoscopes.

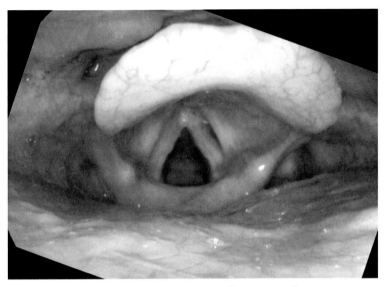

HD picture taken from a typical position above the epiglottis. There is a subtle irregularity to the vocal cord edges, but very difficult to discern much beyond this.

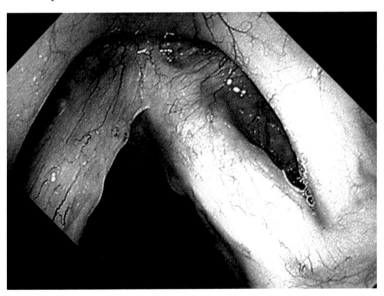

Getting closer and turning on false color imaging changes the view entirely. Capillary dots of individual papillomas can be seen on the left vocal cord.

Data capture

Recording directly to a hard drive is time efficient, and if backups are made in a timely fashion, is probably as robust an archival medium as digital tape. Furthermore, digital hard drive storage certainly saves time when comparing new images to old. I used to have two digital tape recorders and would consume more than a minute sometimes queuing up a new and an old exam on two recorders and flipping the monitor back and forth between recorders for comparisons. If I were building a new lab, I would go with a computer and as large a hard drive as I could afford, preferably a RAID drive.

Database

If you are going to learn much from laryngology, you need a way to retrieve old images, whether digital or analog. I maintain a spreadsheet with a number of parameters recorded on each exam. The essentials are name, date of birth, start and stop times for videotapes and some type of categorization and subcategorizing of disease (see "Part II – Types of voice disorders" on page 85). I keep a note of what was particularly well demonstrated on each exam as well as a rating of the quality of the recording or demonstration of pathology on each exam. A searchable database of this type allows you not only to compare a patient's earlier exam to a current exam, but when I encounter a laryngeal finding that is new to me, I find it invaluable to look up some previous exams for comparison or to answer a question. If you want to give a talk, the rating scale and the brief note saves an immense amount of time compared to a reliance on one's memory.

Presently, I use both a spreadsheet program to keep track of data as well as Apple's Final Cut Pro program. My video files are stored as .mov files on RAID 10 disks, so there is an immediate duplicate made of all video files. Additional copies of all video files are made later, then stored off site.

Light Source

While xenon seems to be the premium light source and halogen is the second brightest, LED bulbs are entering the endoscopy market. A replacement halogen bulb is in the price range of $5 and a replacement xenon bulb is in the range of $600 to $900. I have not priced the LED bulbs. In some systems there is more than one type of light and their may be a very different color balance. It is then helpful to have a camera that can switch at the touch of a button between two white balance settings. Even with white balance, there will be color differences between images. It is very difficult to make color comparisons between images obtained with different endoscopes, with different light sources and from different laboratories.

Bright light is good as long as there is not so much heat that cables melt or people are burned when touching the light cable. If images become too bright as one nears the vocal cords, the video cable can be gradually withdrawn from the light source.

Stroboscope

A stroboscope detects the pitch of the voice, then flashes a shutter at a rate slightly offset from the pitch of the voice. This slightly lower or higher rate of speed records an image creating the illusion of slow motion. With the assumption that the vocal cords are vibrating regularly, the recorded image represents the actual motion of the vocal cords, which otherwise cannot be perceived. Typical NTSC video records at about 29 images per second and a male when speaking vibrates the vocal cords about 100 times per second. Thus, in the case of a constant light source, the edges of the vocal folds move through three cycles for every video frame and consequently appear blurred. Different companies use various techniques for shuttering the light source or shuttering the camera to create a stroboscopic image.

Topical anesthesia

I consider a bottle of topical anesthesia along with some type of applicator to apply the medication to the vocal cords an essential part of any voice lab. It is the poor man's high-definition endoscope. I use four percent lidocaine and most often a curved cannula to drip it on the vocal cords. Two percent lidocaine works as well. I avoid benzocaine as it seems very uncomfortable when sprayed. A sprayer called an atomizer may be used. Lidocaine can even be placed through the cricothyroid membrane.

The laryngologist who can and frequently does anesthetize the vocal cords will offer patients many more answers than those who do not. This is the **most inexpensive** and the **most neglected** piece of equipment in a voice lab.

Duration of the exam

In a new patient, I typically record the vocal cords for about four and a half minutes. The recording time ranges from about three to about nine minutes depending on how easy or difficult it is to find and define the problem. It is uncommon for me to define the cause of hoarseness in an endoscopy much shorter than this. A short video recording by an examiner could be a hint that the endoscopy is just a brief look around for lump or a bump or a color change by the examiner rather than a defining look at how the vocal cords come together and vibrate at different pitches, eliciting any vocal cord gaps and asymmetries.

This makes the duration of the office visit much longer since I often need to review the video's different parts in slow motion, not to mention the amount of time it takes to collect a good history, to perform the vocal capabilities testing and then there is the time afterwards spent discussing the findings with the patient. When I add all this up, I am frequently with a new patient for 45 to 60 minutes.

Summary

I use endoscopy to confirm or disprove what my ears tell me should be the problem. I consider the three-part laryngeal exam:

1. History,

2. Vocal capabilities and

3. Laryngeal endoscopy (with stroboscopy)

to be both an adequate and a complete examination of the patient with a hoarse voice.

Part IV – Parting thoughts

Reflux laryngitis

The Emperor's New Clothes

I would love to lay to rest what I perceive as the greatest mystery in otolaryngology: is hoarseness caused by reflux? Is acid or something else, perhaps something invisible moving from the stomach back up into the throat – refluxing? Chances are, if you have been hoarse and went to see the doctor about it, he probably prescribed some medication for your stomach. Yet, for me, there seems to be a scientific gap in the reasoning behind this nearly ubiquitous diagnosis of *reflux laryngitis* (implying reflux induced hoarseness). No one has ever made any connection between the presumed reflux you are suffering from and vocal cord vibration. Somehow a story was started and it spread with viral effectiveness. Doctors have come to *believe* in "acid reflux" as a cause of most laryngeal disorders.

It is possible that in nearly 10,000 patient exams I have missed every case of reflux laryngitis. It is possible that because treatment of hoarseness with acid reflux medication is so prevalent and as most of my patients have already seen other doctors that I will never see a case of reflux laryngitis. They have all been treated. It is possible that reflux causes swelling of the vocal cords, which should yield a drop in the pitch in people with the condition. I cannot prove beyond doubt that reflux laryngitis doesn't exist, just because I have never seen it.

However, my skepticism comes from college physics. I recall a class on vibrating strings and how a change in mass or a change in tension will change the rate of vibration and consequently change the pitch. I cannot ever recall anyone dripping acid onto a vibrating string and deriving a pitch change out of it.

So until someone can show me a patient where: acid actually touches the vocal cord, and it touches the cord without causing a tremendous spasm, and it then changes the mass or the tension on the vocal cord or somehow alters the airflow between the vocal cords, I remain heartily skeptical of acid reflux (either as blatant as heartburn or the proposed completely silent type) as an explanation for hoarseness.

An article[17] was published in the early 1990s stating that several voice disorders, such as vocal process granulomas, which were seemingly of unknown etiology, might plausibly be from irritation caused by the refluxing of acid from the stomach. Granulomas are found at the back of the vocal cord, the part somewhat close to the esophagus. They are composed of tissue attributed to inflammatory processes. The explanation seemed believable. Dr. Jamie Koufman, the article's author, went on to propose that perhaps other disorders and lesions of the larynx might also be from reflux. Dr. Koufman began lecturing at many meetings promoting the concept of voice disorders being caused by reflux and terms such as GERD (gastroesophageal reflux disease) and LPR (laryngopharyngeal reflux) caught on. Several prominent laryngologists began to promote this concept as well and many, actually almost everyone, have come to believe it.

Many of the studies presented at meetings or published in papers are looking for reflux as a cause of throat symptoms – symptoms such as throat pain, lump in throat, hoarseness and swallowing trouble are lumped together. The presenter tries to associate this collection of general throat symptoms with treatment outcome –

---

17 Koufman, JA. The otolaryngologic manifestations of gastroesophageal reflux disease (GERD): a clinical investigation of 225 patients using ambulatory 24-hour pH monitoring and an experimental investigation of the role of acid and pepsin in the development of laryngeal injury. Laryngoscope. 1991 Apr;101(4 Pt 2 Suppl 53):1-78. This article lays the foundation for the global belief in GERD by saying that reflux *could* cause a number of sensations and conditions in the throat. That *could* has morphed into *does* over the ensuing years in the minds and articles of many physicians.

"less hoarseness." But hoarseness is not a disease. Hoarseness is not a collection of symptoms. As we have learned in this book, hoarseness is either air leak or the production of more than one pitch simultaneously or both.

As a generalization, and it is only a generalization, otolaryngologists frequently have a stereotypical surgeon mind set. Surgeons like to see something and remove it. A lump on the skin can be cut away. A shadow on an x-ray becomes a mass to be excised on the operating table. The professors of otolaryngology select students who think in a similar fashion – *Look! See a mass, cut it out.* This thinking can fail with the voice where the lack of something – the unwanted escape of air – cannot be seen in the same way as a mass can be seen. The laryngologist has to visualize the invisible air.

An otolaryngologist goes to a meeting and the experts say that reflux causes voice disorders. The most famous experts in the profession write textbooks that emphasize reflux as the cause of voice disorders. Almost every otolaryngologist has come to believe that if you cannot see a nodule, a polyp or a tumor, then *ipso facto,* a hoarse voice represents reflux laryngitis. If the patients counters that they do not have any heartburn, the surgeon replies it is "silent reflux." Since all mucosa is red (look in your mouth), the physician has a difficult-to-dispute logic; he sees red. He explains to the patient that red represents inflammation. So, inflammation of the larynx must be from reflux.

Yet, even if experts could somehow see a connection between poor vocal cord vibration and reflux on their stroboscopy equipment, in the day-to-day world of the general otolaryngologist, the diagnoses of GERD and LPR laryngitis have become *de facto,* the wastebasket diagnosis of hoarseness. The general otolaryngologist doesn't see something, so the issue is a silent vocal cord killer – reflux. As mentioned earlier, the popularity of this diagnosis has trickled down to primary care physicians as well.

Secretions and drugs

If we are to accept that air leak and asymmetric vibration are hoarseness, then we are going to have to search for an explanation where reflux causes air leak or asymmetric vibrations. However, there might be an explanation as to why so many people seemingly get temporarily better on medication designed to suppress acid production. The benefit might not be related to suppression of acid production at all, yet might still be related to the drugs used, just another property of the drugs.

I met a baritone, Luigi Senatori, who swore that anti-reflux medication helped his singing. We ran a little single-blind study on a single subject – Luigi. Luigi would come in for an examination and I would not know whether he was taking any anti-reflux medication or not for the previous two weeks. I examined him. He then did the opposite for two weeks and I re-examined him. He switched back to the first condition and I examined him again. He then revealed to me what he had done for each of the two week periods. The major and quite significant difference was that his secretions were very thick and sticky when he was not taking any anti-reflux medication. When he took ranitidine (Zantac), his secretions were much thinner and did not tend to accumulate on his vocal cords. When he was not taking any medication, his secretions were thick and sticky.

Secretions seem to be necessary for vibration. Thin secretions seem to benefit vibration. When I examine normal patients, clear, thin secretions tend to ride on the superior or top surface of the vocal cord. The vocal cord vibrations appear supple, resembling a wave washing over the vocal cords viewed during stroboscopy. Patients with very dry vocal cords have a great deal of difficulty getting any vibration out of the vocal cords. Patients with thick secretions have a wave on the surface of the vocal cord, but tend to accumulate the secretions between the vocal cords that then interfere with vibration to some degree until cleared away.

Usually secretions, thick or thin, are most notable when there is an irregularity of vibration. They tend to accumulate at the point of irregular vibration. For example, whenever there is a swelling along the edge of the vocal cord, the vibrations are dampened at that location from mass and stiffness, leading mucus, thick or thin, to accumulate on the elevated or stiff area. If enough mucus accumulates, it can be audible. Mostly the individual has the sensation of excess mucus and then tends to clear their throat. In another situation, when the vocal cords do not come completely together (bowing, paresis), mucus tends to accumulate at the closest junction of the vocal cords. Even in patients who complain of excessive post-nasal drip, the issue tends to be mucous buildup somewhere along the vocal cords rather than actual excessive or dripping mucus.

I will speculate that since ranitidine is an histamine blocker, there might be some blockade to the production of thicker mucus. Perhaps some property of the drug stimulates the serous glands (producer of thin, watery secretions) or blocks the mucous glands (producer of thick, sticky secretions). This effect from the drug may be entirely unrelated to the blockade of acid production in the stomach.

My suspicion is that if secretions are made thinner, there will be less tendency for secretions to stick to any pathology of the vocal cords. The patient will have the sense that their voice has improved (somewhat) and the sense that their post-nasal drip has improved, even if the underlying problem (vocal cord pathology) has not changed. He or she will feel better, but never quite completely cured unless the only problem is truly thick secretions. Luigi seems to have only thick secretions interfering with his voice. He is happy taking the anti-reflux medication.

The color red does not vibrate (at least at an audible frequency). The color red does not affect vibration of the vocal cords, so cannot impact either air leak or diplophonia. Consequently, redness is not a cause of hoarseness. It seems reasonable to me that consumers of health care ask their physicians for a reasonable explanation rather than a pill for the color red.

287

Value

How much should you pay?

Mrs. Seeka Tris comes in with her daughter Susan for a second opinion. They really like their doctor, and don't want him to know that they are seeing me, but it has been such a rough time that Susan thought it might be worthwhile to get her mother a second opinion.

Susan says to me, "Mom had mitral valve surgery last spring. When she woke up after the surgery, things didn't go very well. She ended up in the ICU for 10 days with a tube in her throat on a ventilator. She recovered, and we're happy, but when the tube came out she couldn't speak except for a whisper. The cardiac surgeon counseled us that this happens from time to time, but it will likely recover, lightly admonishing her to be a patient patient."

Mrs. Tris went home with a soft voice. Over about a month, her surgeon's prediction proved to be correct and her voice began to recover. By two months, she was speaking much better, almost with a normal volume. The surgery seemed to have taken a lot out of her though, as she was chronically short of breath. She suspected that must not be unusual for heart surgery. She was happy to have survived. However, her breathing worsened enough one night that she called her daughter. Together they headed to the emergency room. (This is such a typical story: hoarseness after having a tube in the throat followed several months later by an improving voice, but worsening breathing.)

At the emergency room, she is breathing quite rapidly and noisily. The ER physician runs quite a few tests including a ventilation-perfusion scan and tells her that she should be relieved. She does not have a pulmonary embolus – a serious, life threatening condition with a blot clot in her lung. She can go home tonight. She suggests that Mrs. Tris see a pulmonologist – a lung doctor.

In a few weeks, she visits the pulmonologist who orders some breathing tests, performs a bronchoscopy. After, he tells the mother-daughter duo that the problem is with her vocal cords. She is referred to an ENT doctor.

A couple of weeks later, the ENT doctor looks in. She tells Mrs. Tris that her vocal cords are not working well and she needs to see a subspecialist, a laryngologist. Another appointment is made.

The laryngologist puts a scope into her nose to view the vocal cords and makes a diagnosis of bilateral vocal cord paralysis. He orders an EMG study; an electrical test of the muscles of the vocal cord to see if the nerves have been injured.

She makes an appointment with the neurolaryngologist for the EMG. There, needles are passed through her neck, into her vocal cords and she is told that the signals are abnormal. He believes the EMG confirms that she does have paralyzed vocal cords. He suggests a botulinum toxin injection into one of the vocal cords to weaken it and allow it to open up and improve her breathing.

Six months have now passed since she woke up without a voice. She followed through and had the botulinum toxin shot two weeks ago and her breathing is a little worse after the injection rather than better. A little frustrated, this is how she ended up today in my office for another opinion.

For me, with a history of loss of voice after a tube in the throat, whether the tube is in for two hours, two days or two weeks, there are two likely reasons for the voice loss. Let's consider what the most likely reasons are.

Tubes are placed into the throat for breathing. They are typically made of a semi-rigid plastic with a soft, inflatable cuff near the end that can be filled to create an air seal. If the cuff is inflated too much and too close to the undersurface of the vocal cords, it may put pressure on the nerves as they enter the vocal cord muscles. If blood stops flowing to the nerves for a period of time, they become injured and stop transmitting a signal, so the vocal cords will move less or will not move at all after the injury. If the injury is severe enough, the nerve cells die. Consequently, one consideration for the physician is nerve damage.

The other typical problem is from the hard part of the plastic tube resting between the vocal cords. The tube typically places the most pressure on the cartilages between the vocal cords at the back of the larynx. Even in as little as two hours, the pressure may injure the mucosa. Several days after the injury ulcers form. After the tube is removed, the ulcerated area contracts and scars over several weeks, pulling the back of the vocal cords together. Scarring is the other consideration for injury after a tube is in the throat.

In a nerve injury, the vocal cords will not be moving well after the tube is removed. However, this nerve (the recurrent laryngeal nerve) has a very strong propensity to regrow and over a few weeks to a few months the nerve sprouts new endings that grow back to the muscles. The individual fibers may grow back to the same muscle, but at times grow back to a different laryngeal muscle and this causes the vocal cords to move, but often in an inappropriate way. So several months later, the voice may become strong again, but the vocal cords may fail to open during breathing and so the person has a reasonable voice, but makes noise when the air passes through the minimally open vocal cords.

If Mrs. Tris had mucosal ulcers from the tube putting pressure between the vocal cords, then she may initially have a lot of swelling preventing the vocal cords moving together to vibrate. As the swelling goes down, and as the ulcer heals, the tissue contracts over the ulcer and pulls the vocal cords closer together. Her voice gets stronger, but subsequently the breathing gradually becomes more difficult and more noisy.

When I first look in her throat with the endoscope, her vocal cords do not open very much. In a broad sense, I could say that she has a bilateral paralysis. However, this endoscopic overview will not distinguish between the two possibilities above.

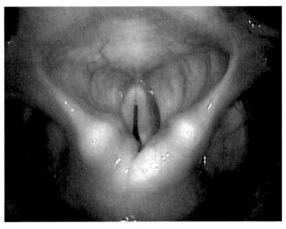

Flexible endoscopic view of Mrs. Tris' vocal cords during breathing in. This is her maximal opening.

With these two possibilities in mind – paralysis or scarring with fixation – I drip some numbing medication on to her vocal cords and pass the endoscope under the arytenoids at the back of the larynx. I place the endoscope up against the back of her vocal cords looking at the area where the endotracheal tube would have been sitting for 10 days. I find a band of scar tissue just beneath the vocal cords, holding the vocal cords together. At most, she can open her vocal cords about 2 millimeters because of this scar band.

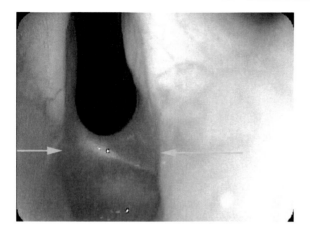

Ultra close-up, enlarged view of the back of the vocal cords (hidden underneath the arytenoids in the previous photo) where a web is running between the vocal cords (arrows) preventing them from opening. The white areas at the top of the photo are the vocal processes that open and close the vocal cords.

I think back on the cost of Mrs. Tris' diagnosis: the emergency room visits, the tests, the procedures, the consults and not the least, the amount of time it took from her life for her to get the correct diagnosis. It is quite possible to go in with surgery to divide this scar tissue, but only if you know what the problem is. At times the joints will need to be mobilized as they have been stuck in one position for many months and the longer that delay causes them to sit in a fixed position, the less likely it will be to get them moving again.

Ultimately, when you visit a doctor about a problem, don't ask yourself, "How much is this going to cost?" What you really want to know is, "How much value will I receive for this advice?" This is a conundrum in medicine. Most times you have no idea how valuable a doctor's advice is until you finish the treatment and it worked or it didn't. Almost any price I charge is acceptable to patients – if they get better. People attach great value to their health, at least once they are ill. No matter how little I charge, if a patient doesn't improve, I am too expensive.

293

Interestingly enough, our body functions like a well engineered system, and problems with the system are generally quite understandable in scientific or engineering terms. Sound production and airflow between the vocal cords is particularly easy to understand. If a physician has the luxury of good endoscopes and a recording system, the physician should not only be able to diagnose a problem, he should be able to explain to the patient why it is not working. With video, he should even be able to clearly show the patient what is working and what is not working. The physician's explanation should make sense to anyone. For the voice, the explanation should include an air leak or an asymmetric vibration as the cause and the patients should be able to see this. For breathing problems of the larynx, the explanation should demonstrate how air flow is limited by the larynx.

If the explanation a doctor offers seems mystical or mysterious in any way and the patient jumps to the conclusion that the doctor must be brilliant, it is likely fraught with pitfalls. Baffled by an explanation, a patient is at least possibly correct when the thought enters her mind, "I am not sure my doctor knows what he is talking about." A weakness of some physicians is they cannot admit when they don't really know what is going on. They feel the need to provide an explanation every time. If this explanation doesn't make sense, there is a high probability that it is a contrived explanation.

Mysteriously, some patients will improve with the incorrect treatment. There is often a significant placebo effect. There are also many conditions that the body rectifies spontaneously and this spontaneous improvement may incorrectly be attributed to the physician's ministrations. Most odd to me is that so many people want their individual physician to be correct, that any improvement is attributed to the doctor's intervention and yet any deterioration is attributed to themselves or some unknown external factor.

So, caveat emptor.

Errors & learning

Waiting and observing

I'm fairly certain that before I devoted myself to the study of the larynx, I placed a tracheostomy in a patient with a nonorganic loss of voice with stridor (noisy breathing). I put the tracheostomy in with the patient's permission and she felt very relieved, which I thought confirmed my excellent decision-making. Her breathing problem was gone and her voice was recovered the next day, though she now had a plastic tube in her neck. Since, in retrospect, I now understand the natural course of nonorganic stridor, I presently treat similar patients much more efficiently and cost effectively in the office with gentleness, patience and persuasion.

Over time in my practice, a finite number of voice problems appear over and over. The study of voice seemed to be getting simpler and simpler as hoarseness seemed to boil down to only two types of problems – air leak and vocal cord asymmetry. Laryngology seems less and less complicated.

The less an examiner knows about voice, the more distracted they become by everything they see. The more an examiner understands voice, the simpler the examination of the hoarse voice and the more focused the examiner can be on identifying the problem. The examiner is less likely to harm someone when he understands that a nonorganic stridor doesn't require the expense of hospitalizations, consults, asthma drugs and surgeries like tracheostomy. I removed her tracheostomy after two days and her voice was restored and her breathing normal, even if there may have been a better way.

I have been treating otolaryngology problems since my otolaryngology residency began more than 25 years ago. For more than half that time, I have been seeing only patients with laryngeal problems. Each laryngology patient's symptoms and exam are a puzzle to be put together. The puzzles have become more simple for me to understand, yet I still continue to learn something new by observing my patients.

Sarah S. had a procedure where I removed the front portion of the voice box and shortened the vocal cords. I perform this surgery in persons whose vocal pitch is too low for their appearance as mentioned earlier in this book. The surgery raises the comfortable speaking pitch of the person and is frequently used for genetic males who are transitioning to female. They were exposed to testosterone and have too low of a vocal pitch for their appearance. It is a significant surgery and should there be significant swelling after surgery, breathing can be impaired.

Sarah S. came back for a check up the day after her surgery saying that she felt short of breath during the night. She wasn't making any noise breathing, but definitely felt that something was not right. I sprayed some numbing medication in her nose to look at her vocal cords (which I almost always do in everyone before an endoscopy to make the procedure comfortable). The vocal cords were not swollen, but it was immediately obvious that the structures above the vocal cords – the arytenoids – were very swollen and bruised. They leaned in towards the vocal cords and moved in and out towards the vocal cords with each breath. The swollen arytenoids were narrowing the opening for breathing more than 75 percent.

The problem might seem like an obvious one from the above description. I knew she was breathing fine before surgery. I narrowed her breathing opening and then post-surgery swelling blocks 75 percent of her airway. Yet, my experience suggested that the problem was not due to the obvious visible swelling. She pointed out that she felt even more short of breath after I sprayed her nose with the anesthetic and I know that quite frequently the anesthetic runs down the throat and gives patients the sense that they cannot breathe even though they can. My medication was a possible contributing factor.

The visually obvious answer did not work for me because I see patients with both vocal cords paralyzed who get about their life pretty well with about 95% of their visual airway blocked. Usually people do not have trouble from narrowing of the airway until they are making noise with every breath and Sarah was not making a single sound while breathing. The arytenoids were not being pulled into the airway. They were not vibrating with each breath, so even though the swelling was moderately impressive visually, her sensation was out of proportion to her exam. Based on my experience, I couldn't put all the pieces of her puzzle together.

I asked her to wait in the office, perhaps her symptoms would change after the numbing medication wore off. A half hour later she coughed and up came a blood clot. She felt much better. We took a look again at her larynx with the camera through the nose. The vocal cords and the swelling above them were just the same as before, even though she felt that she could breathe much better, essentially normally now.

So now the puzzle could be put together. My surgery was the ultimate cause of the problem. Some blood from the incision must have run down into her lungs and I had asked her not to cough so as not to bother the sutures that I had placed into her newly tightened vocal cords. She had followed my instructions completely. But the blood running down her windpipe, had clotted and blocked off some portion of her lungs, reducing her breathing ability.

I learned that just because something looks abnormal – her swollen arytenoids – looks bad or even looks like it might be causing the problem, the seemingly obvious answer may not be the correct one. I already knew that the swelling, even though it looked bad, shouldn't make her feel as bad as she did. I knew that the problem was from the surgery as far as timing was concerned. I just couldn't put these two issues together.

There is an initial tendency as a surgeon to feel guilt for causing a problem and wanting to rectify it quickly, perhaps believing the patient will have more confidence in me. I noticed this problem even more in academic settings where several residents might be around, there might be other physicians nearby, a nurse and therapist may be on rounds as well. This can drive the physician's need to be recognized as an authority to hastily reach a quick (and perhaps incorrect) decision. Sometimes sitting back and watching the problem evolve with judicious reassessments, steers one in the right direction. Careful watching and judicious waiting both solved the problem and taught me something.

Models of diagnosis

The vibratory model

No doubt there are still disorders of the larynx that I have not yet seen, and there are likely more variations of the disorders I have described in the previous chapters. So it will not surprise me that this book will need revision. In particular, I am sure there are other types of tumors I have not seen. There are certainly other configurations of neurologic disorders. However, by understanding that hoarseness is caused by air leak or asymmetry of vocal cord vibration, I will already understand a part of the disease when I see a new pattern of either, even before I can apply a name to it. Mind the gap.

The color model

Webster Hess, a young general otolaryngologist, took a look at some vocal cords and saw a white lesion. He said to his patient John White, who was complaining of hoarseness, "You have leukoplakia on your vocal cords. It could be cancerous. We should take it out." John agreed. It seemed prudent to get out something that could be cancer.

On the biopsy, there was no cancer to be found, only dysplasia. John was relieved, but John's voice was worse after the surgery. Several months later, Webster removed more leukoplakia, again benign. Again John's voice was a bit worse. When I met John two years later, he had been to six doctors for his hoarseness, had several different treatments and recently another biopsy, this time of the false vocal cord that showed cancer.

I had a chance to speak with Webster. He said, "I was taught that anytime there is a lesion near the vocal cord's anterior commissure, it is difficult to get a good voice after surgery, so his hoarseness after surgery didn't surprise me."

So where is the problem? Webster did what he was taught – surgically cut out something of abnormal appearance. The pathology was benign at the time. The present problem is that John will need a larger surgery or more aggressive treatment now than he would have needed if the cancer had been discovered several years ago. The cancer was there, causing the hoarseness and continuing to grow while John hunted for an answer.

Webster looked for a lesion to match with the hoarseness. On endoscopy he saw white on one of the vocal cords, so he presumed the white patch, the leukoplakia was causing the hoarseness. He did not think in terms of whether there was an air leak or an asymmetric vibration. He focused on the abnormal white color. When John didn't improve after removal of the leukoplakia, Webster didn't ask himself, "What am I missing?" Rather he told himself, "I was taught that sometimes hoarseness from leukoplakia near the front of the vocal cords doesn't get better even after surgical removal."

But we know that color doesn't vibrate, so white, red or even green vocal cords would have no bearing on a hoarse vocal quality. But if your method of diagnosis depends on color, you risk completely missing the diagnosis and not knowing why.

It this case, much of the tumor was within the false vocal cord. There was also some tumor pushing from the deep aspect of the true vocal cord, creating a convex edge to the true vocal cord. Leukoplakia can make the edge of one vocal cord stiffer or heavier than the other causing an asymmetry between the vocal cords. These represent at least three potential reasons for John's hoarseness. The pressure from the false vocal cord rubbing and dampening the vibrations on one of the true vocal cords could cause an asymmetric vibration between the vocal cords – rough hoarseness. The convexity of one true vocal cord could cause the hoarseness by creating a gap or gaps on either side of the prominence – husky hoarseness. A normal vocal cord wrapping around a convex vocal cord will not vibrate and so if they were held apart such that they touched only at the most convex point, the good vocal cord would flutter – rough hoarseness. A vocal cord that is slightly stiffer than the other may vibrate at a second pitch creating diplophonia – rough hoarseness. If John learned to compensate for the convex vocal cord by not closing his vocal cords all the way to avoid flutter, he could leave a large opening at the back of the vocal cords which leaked air – husky hoarseness.

Webster did not record his examination so I cannot go back in time and determine what was the initial type of hoarseness. I can say that a precise diagnosis would orient Webster to the surface of the vocal cord, to the curved margin of the vocal cord or to the false vocal cord depending on what the findings were. He would then know whether to remove the leukoplakia to find the problem or whether to look at or biopsy other parts of the larynx. He would know whether to order a CAT scan to look under the surface of the false or true vocal cords. He would not be distracted by the color of the vocal cord and even if initially he was distracted, he would look again more closely if the hoarseness did not resolve after the removal of the leukoplakia.

I would be wary of the color model of diagnosis. Color does not alter pitch or volume. It basically alters nothing at all about vocal cord vibration and so bears no cause and effect relationship with hoarseness. As a physician, Webster probably doesn't want to miss a second cancer. One is enough.

The reader

How can you use this information?

This book is by no means an exhaustive compendium of voice disorders, though it does propose that all hoarseness can be explained either in terms of air leak or an asymmetry of the vocal cords. For more information on voice disorders, updates to the book, additional photos, videos, finding a laryngologist or other voice information, please feel free to visit www.voicedoctor.net.

The larynx contains a valve, the vocal cords, that affect the voice, breathing and swallowing. Laryngeal structures other than the vocal cords can also be affected, so there are problems of the larynx that might not cause hoarseness. However, most of the problems with the larynx are amenable to discovery with a very similar examination as described in this book. So if you have a problem in your throat, seek out a detailed exam.

I suggest that "silent reflux," GERD and LPR as causes of hoarseness represent misunderstandings of the role of mechanics in vocal cord vibration. While I believe this misunderstanding on the part of physicians is one of omission and not likely intentional, I hope that if patients cajole physicians to provide a reasonable explanation, physicians will seek a better understanding of vocal mechanics. I find it less and less excusable for physicians, presumably scientists, to "believe in reflux" (and unflinchingly prescribe a pill for a mechanical problem).

Even when I follow up with a patient after a treatment, I always leave open to the patient the possibility that my treatment didn't work. I ask, "Did the treatment work? Or are you the same? Or have you gotten worse with my suggested treatment?" I am well aware of the placebo effect of treatment, as well as the patient's unfortunate desire to please me, as if pleasing me mattered. Both the placebo effect and the patient's desire to please their physician work to lead the patient to say they are better, even if they truly might not be improved. I try to give permission to my patients the opportunity to say that I was incorrect in my diagnosis. Don't try to please your physician. That is not what you are paying for. Give him an honest appraisal of what worked and what didn't.

When you are in the office, go ahead and let your voice sound bad, especially you – the singer! This is not the time to apologize and attribute your vocal problem to bad technique or not being warmed up. This is not an audition. Whatever sound comes out, let it come out. Maybe your problem is technique, but maybe the issue is a physical impairment of vibration. Don't try to compensate. Compensation is the body's natural reaction to a bad voice. In fact, a laryngologist's principal diagnostic tool during an endoscopy is to remove compensation and expose the vibrating problem. A video recording while the voice is rough is very helpful.

Don't be afraid to question your doctor. He is not infallible. Does your doctor's explanation make sense? If your doctor isn't reasonably willing to answer your questions in an understandable manner, seek care elsewhere. Just because he commands a high price, doesn't mean you cannot seek high value from his explanation and treatment.

If you are a physician, I hope this book offers you a plausible diagnostic algorithm for voice disorders. Even if you don't have an endoscope, you now have a clinical way of thinking about voice disorders. You may develop an ability to hear husky and rough hoarseness.

If you are prescribing anti-reflux medication for hoarseness, make the trial a reasonably short one. I hope you will also see the value in referrals to physicians who specialize in laryngology.

Reasonable expectations

Did your voice doctor offer you a good opinion?

Did your laryngologist make a video? Did he explain the problem to you in a way that you understand the video? If he didn't explain your hoarseness to your satisfaction, get a copy of the video and take it home.

First, identify the vocal cords in your video. Watch them come together on the stroboscopy portion of the exam. As they vibrate, can you identify any persistent gap between them that never closes. That should explain huskiness in your voice.

Then, look for any asymmetry between them. The vibrating cords should be mirror images of each other. Any asymmetry is a potential explanation of roughness. Look at the portion of the video when your voice sounds bad. If no recording was made at a pitch or volume that brought out your hoarseness, then the problem could well have been missed. If you are still in the office with your laryngologist, don't try to sound good. He will gain the most information by recording video under the conditions when your voice is sounding hoarse.

If the air leak or asymmetry is not obvious, did the examiner get close to the vocal cords? When I am close to the vocal cords, even though they are less than an inch long, they often no longer fit within the video screen. That is how close a laryngologist can get. The vocal cords can more than fill the screen.

If the air leak or asymmetry is not obvious, did the examiner make a recording at high and low pitches and at loud and soft volumes? The more ways your vocal cords are tested the more likely that an accurate explanation for hoarseness can be found.

These are all ways to assess how adequate is your laryngeal examination and how you are going to get that frog out of your throat.

Glossary

abduction: the vocal cords move away (AB: Latin for away from) from each other.

adduction: the vocal cords move toward (AD: Latin for toward) each other.

Adam's apple: the protruding top center portion of the thyroid cartilage. It forms after exposure of the cartilage to testosterone.

arytenoid cartilage: a pair of cartilages that rotate the vocal cords open and closed. They are the primary component of the bumps easily visible when looking at the back of the larynx and singers often initially mistake these bumps for vocal nodules.

aryepiglottic fold: the ridge of mucosa between the epiglottis at the front and the arytenoids at the back that acts a bit like a dam to keep food away from the vocal cords.

cricoarytenoid joint: the arytenoid cartilages glide over the surface of the cricoid cartilage to open and close the vocal cords.

cricoid cartilage: the ring-like cartilage that is the foundation of the larynx. It is usually the most prominent bump on the front of a female's neck, but is the smaller bump on the male's neck.

cricothyroid joint: the cricoid and thyroid cartilages are connected by a joint on each side at the bottom back of the thyroid cartilage. Contraction of this joint allows production of higher pitches and louder sounds.

cricothyroid muscle: the muscle that puts our voice into falsetto. It is located on the outside of the larynx, just under the skin.

cricothyroid space: a groove at the front of the neck, just below the Adam's apple, somewhat famous before the Heimlich maneuver. It is the closest the windpipe comes to the surface of the neck and a person choking on something could often breathe again if a knife was put through this space. It still works, but most people seem to prefer the Heimlich maneuver. This space narrows when a person sings in falsetto.

dystonia: originally thought to be a disorder of muscle tone, it really represents an inappropriate degree or timing contraction of muscles during intentional movement.

epiglottis: a soft cartilage that folds over the larynx during swallowing directing food and water away from the lungs.

esophagus: the tube through which food and water passes. The start of the tube is at the back of the larynx and is closed during rest and opens very, very briefly during swallowing.

interarytenoid muscle: the muscle that keeps the vocal cords together during prolonged phonation.

larynx: the organ that acts as a valve connecting the throat to the trachea, diverting food into the esophagus, directing air to the lungs and generating sound.

lateral cricoarytenoid muscle: the muscle that brings the vocal cords together. It contracts partway to keep the lungs filled with air. It contracts fully or almost fully during voice production, coughing or swallowing.

pachydermia: a rugged appearance of the mucosa, usually between the arytenoids where it is loosely attached, such that it resembles an elephant's skin.

phonation: the production of sound or voice.

posterior cricoarytenoid muscle: the muscle that opens our vocal cords for breathing or sniffing.

presbyphonia: aged person's voice.

reflux: the idea that the contents from the esophagus or stomach travel in reverse. The general thought is that the contents of the stomach are acidic and when acid travels in reverse up the esophagus, it may do some harm.

reflux laryngitis: The larynx is described as red and inflamed, presumably as a result of acid touching it.

Reinke's edema: a jelly-like fluid that expands Reinke's space in chatty smokers.

Reinke's space: the lubricating layer in the vocal cords, named after the man who first injected the area with fluid to see how big it was.

thyroarytenoid muscle: the muscle within the vocal cord, also called the vocalis, that raises the pitch of the voice when contracted.

thyroid cartilage: the large shield-like cartilage at the front top portion of the larynx.

tumor: cells that grow without stopping. If they tend to stay in one place, they are typically called benign. If they tend to break off and spread to other parts of the body, we call them malignant or cancer.

vocal cords: a pair of soft, straight pieces of tissue, attached at the front to form a V-shaped opening, which close like a valve, yet are flexible enough to vibrate. They consist of a ligament, a muscle (thyroarytenoid) to tune them, a lubricating layer and loosely attached mucosa to vibrate and create sound.

Index

W

X

Z